Meaningful Games

MEANINGFUL GAMES

Exploring Language with Game Theory

Robin Clark

The MIT Press
Cambridge, Massachusetts
London, England

© 2012 Massachusetts Institute of Technology

All rights reserved. No part of this book may be reproduced in any form by any electronic or mechanical means (including photocopying, recording, or information storage and retrieval) without permission in writing from the publisher.

For information about special quantity discounts, please email special_sales@mitpress.mit.edu.

This book was set in Syntax and Times New Roman on 3B2 by Asco Typesetters, Hong Kong. Printed and bound in the United States of America.

Library of Congress Cataloging-in-Publication Data

Clark, Robin Lee, 1957–
Meaningful games : exploring language with game theory / Robin Clark.
  p. cm.
Includes bibliographical references and index.
ISBN 978-0-262-01617-9 (alk. paper)
1. Mathematical linguistics. 2. Game theory. 3. Language and logic. I. Title.
P138.C43 2012
401'.43—dc22
                                        2011005883

10 9 8 7 6 5 4 3 2 1

Sous les pavés, la plage!
—Situationist slogan, May 1968

Contents

Preface xi
Acknowledgments xvii

I THE SOCIAL SIDE OF MEANING

1 Platonic Heaven 3
The Puzzle of Reference 3
Use, Mention, and Truth 4
The Language of Thought 7
Concepts, Mentalese, and the Informational Universe 10
Language and the World 12
Platonic Heaven in a Box 13
Inferences and Mentalese 16
Further Reading 18

2 My Fall from Platonic Heaven 21
Phrase Structure Grammar 21
Grammar and Compositionality 23
Thinking and Computing 25
The Heaven in Your Head 28
Brains in SUVs 30
Symbols and Proofs 31
Into the Chinese Room 36
The Social Nature of Intention 38
The Excesses of Youth 39
Further Reading 41

3 Meaning and the Social Contract 43
Choice and Meaning 43
Internal Predicates and External Behavior 44
Public Knowledge 46
The Economics of Meaning 50

Physical Computation and Social Computation   53
The Sociolinguistics of Meaning   55
Further Reading   58

II   GAMES AND TRUTH

4   A Primer on Games   63
The Cake Game   69
Sequential Games and Backward Induction   75
The Holmes-Moriarty Game   80
"Ideal Free" Ducks and Mixed Strategy Nash Equilibria   85
Mixed Strategy Nash Equilibria and Language Variation   88
Coordination Games   89
The Prisoner's Dilemma   93
Cooperation: The Stag Hunt   98
Evolutionary Games   106
Case Marking Systems   112
Further Reading   122

5   A Game Logic for Natural Language   125
The Tale of Abélard and Eloïse   128
Syntax   129
Games and Models   137
Atomic Sentences   137
Negation   140
Logical Connectives   142
The Aristotelian Square of Opposition   150
Prospects   172
Ambiguity   172
Monotonicity   173
Compositionality   174
Limitations   176
Further Reading   177

III   GAMES AND THE WORLD

6   Common Knowledge   181
Coordinated Attack   182
Definite Descriptions and the Mutual Knowledge Paradox   183
Common Knowledge and Bounded Rationality   191
Miscommunication   199
Presuppositions and Accommodation   204
Reconciling the Assumptions   212
Further Reading   212

Contents

7   **Lexical Games**   215
    Games for Finding Words   215
    Orderly Communication and Utility   221
    Playing the Odds   226
    Clues from the Context   229
    Back to Descriptions and Common Knowledge   232
    Equilibrium Selection and Implicature   234
    Down the Garden Path   237
    Further Reading   242

8   **Two Examples: Pronouns and Politeness**   245
    Discourse Pronouns   245
    Politeness, Power, and Implicature   263
    On Game-Theoretic Analysis   279
    Further Reading   280

9   **The Social Ecology of Meaning**   283
    Games and Prototypes   285
    Metrics, Central Tendencies, and Focal Points   289
    Semantic Landscapes and Meaning Niches   301
    Semantic Hierarchies and Defaults   309
    Homophones and Polysemy   314
    Into the Artificial World   321
    Further Reading   327

    Notes   329
    References   333
    Index   345

Preface

I want this book, the one you're holding now,[1] to introduce you to a way of thinking about language that I've found very interesting and helpful. The idea is that we use grammar strategically to signal our intended meanings. By *strategically*, I mean that my choices as a speaker are conditioned by the choices you as a hearer will make in interpreting what I say. In short, I've found game theory—the theory of decision making when the outcome of the decision depends on the choices of others—to be enormously helpful in thinking about a wide variety of linguistic puzzles. Let me try to give you an idea of what I mean by this.

If you take a simple word like *and*, you'll find that it's capable of some quite complex behavior. Certainly, there is the familiar *and* of temporal sequencing: the sentence

(1) They got married and had a baby.

is decidedly different from

(2) They had a baby and got married.

The logician might shudder and point to the following pair of sentences, which are surely equivalent:

(3) a. The House of Representatives has 435 members and the Senate has 100 members.
    b. The Senate has 100 members and the House of Representatives has 435 members.

Both sentences in (3) amount to pretty much the same thing. While order matters in (1) and (2), it seems not to matter in (3), where all that is required is that the sentences on either side of the *and* be true. What about

(4) I added nitrate to the damned thing and it blew up!

Surely, more than temporal sequencing is going on in (4); we might infer that the reason it blew up was that I added nitrate to the damned thing.

Well, fine, we might say, we just need to define three kinds of *and*: one kind for temporal sequencing, another for causation, and a third as the logician's conjunction. Perhaps a little thought will reveal still more kinds of *and*.

Surely, though, we're missing something important by supposing that there are three different kinds of *and*. The treatment is compatible with the idea that there are three different words—one for temporal relations, one for logical relations, and one for causation—which just happen to sound alike, a peculiar accident of English.

We ought to entertain the idea that the three different "meanings" of *and* flow from different uses of the same semantic thing. Would a language have a different word for each of the three uses? Perhaps, but certainly most languages use a single word to serve each of the three different purposes. Something more than mere accident is going on.

The great philosopher of language H. Paul Grice thought that there was more here than mere coincidence. He argued that the regimented semantics of the logician didn't quite capture things. Rather, a different kind of logic was needed, a "natural" logic, that could never be replaced by the logician's regimentation:

> Moreover, while it is no doubt true that the formal devices are especially amenable to systematic treatment by the logician, it remains the case that there are very many inferences and arguments, expressed in natural language and not in terms of these devices, which are nevertheless recognizably valid. So there must be a place for an unsimplified, and so more or less unsystematic, logic of the natural counterparts of these devices; this logic may be aided and guided by the simplified logic of the formal devices but cannot be supplanted by it. Indeed, not only do the two logics differ, but sometimes they come into conflict; rules that hold for a formal device may not hold for its natural counterpart. (Grice 1975, 43)

The idea is a compelling one—a more abstract logic guides our use of language in such a way that meanings emerge. But what kind of logic could it be? As Grice observes, it certainly isn't the formal logic we might learn in a philosophy or math class. It would have to be something prior, something we all share.

We might suppose that Grice's natural logic is really just the rational use of grammar to signal meaning. On this view, given a context, we use the grammar as a tool to signal meaning; the choices arise from our knowledge of the context, our knowledge of grammar, and our communicative intention. On this view, grammar is a tool that we deploy to get things done.[2] Underlying our use of grammar is a logic of rational decision making. In order to get things done, we must make communicative decisions based on the (potential) decisions of our interlocutors.

Game theory explicitly concerns itself with rational decision making when the outcome of the decision depends on choices made by other (rational) agents. It therefore provides a mathematics that allows us to develop theories of the kind of decision making crucial to understanding linguistic behavior.

Linguistics, particularly in North America, has been dominated by two trends that, while initially helpful, have hardened into dogmas. The first trend is solipsism. The proper subject for linguistic theory, according to this line of thought, is grammatical representation, largely divorced from the content of these representations. The focus on grammatical representations led to an explosive growth of linguistic theory during the second half of the twentieth century. Certainly, the data uncovered and classified by this revolution have been crucial to our understanding of linguistic forms and language diversity.

The second trend dominating linguistics has been the conflation of linguistic explanation and grammatical explanation. As we have learned more about language meaning, for example, the tendency has been to make the semantic component—and, more often than not, the syntactic component—of the grammar more complex. Thus, nodes corresponding to pragmatic functions have been added to syntactic trees, and the semantics itself has been rendered more complex in the service of the goal of explaining aspects of language that might better be accounted for in terms of the use of grammar rather than the grammar itself.

Game-theoretic pragmatics runs counter to both trends. Game theory is, by its very nature, antisolipsistic. The solipsism in current linguistics, and in cognitive science more generally, has outlived its usefulness. The only way to properly understand meaning is to grapple with its social nature; language, after all, is the bridge between our private mental worlds and the public world of social interaction. I argue that, in fact, it is the social that gives content to our mental lives.

The idea that use determines meaning is hardly new; its roots lie in the work of Wittgenstein, Austin, Grice, and many others. Happily, game theory gives us a formal language for working out these ideas. The resulting theory of use will allow us to account for many aspects of linguistic meaning, and the grammar itself can be simplified. The resulting theories are nevertheless precise and subject to empirical testing.

One of the pleasant aspects of game theory is that it allows us to unify many aspects of linguistics that seem, at first glance, to be disparate. For example, we can begin to see profound connections between sociolinguistics and the study of meaning. These connections can be followed into neurobiology, as I suggest at various points in the book. The game-theoretic approach to language promises to open connections between behavioral economics, social evolution, and neuroeconomics.

I would go one step further and argue that game theory returns linguistics to the heart of the social sciences. In recent years, game theory has helped pave the way toward a systematic study of the development of conventions, the evolution of altruism, and reciprocal behavior. Language provides a platform to study all these things; the evolution of Gricean implicature is but one instance of the broader evolution of cooperative behavior.

This book is intended as an accessible introduction to game theory and the study of linguistic meaning. I have tried throughout to keep the tone light and to presuppose little specific knowledge; my intention is to make the ideas available to a wide audience. Many of the ideas I touch on are, by their nature, obscure; nevertheless, I believe that discussions about the nature of meaning, meaning as the outcome of strategic reasoning, are vital to a wide audience. I hope that undergraduates, graduate students, and general readers with an interest in language will find something useful here. The time has come for linguists and other cognitive scientists to make these ideas available to a mass audience, lest we become another obscure guild, open only to a few specialists.

In order to make the book as accessible as possible, I have been sparing in footnotes, have left most bibliographic matters for the end of chapters, and have tried to keep the mathematics down to some simple algebra. Formal definitions have been placed in boxes outside the main text so that readers who are not interested in that level of detail can simply pass them by while still reading the main text.

The first part of the book, chapters 1–3, is an extended argument in favor of the social basis of meaning. While a definitive argument is, of course, impossible, I think the weight of evidence strongly supports the social nature of linguistic meaning. I have occasionally wandered into the realm of the memoir, my hope being that I can show why meaning matters so much to me. The issues here, grounded as they are in analytic philosophy, can seem arcane to a nonspecialist; nevertheless, the questions I raise are of general importance. The main arguments in favor of the economic and ecological nature of meaning are in chapter 3. Chapter 9 attempts to work out the nature of the system more precisely. Part I is a prelude to the study of games and meaning; it motivates the use of game theory in the study of language without being about game theory proper.

Part II turns to game theory. Chapter 4 is a brief, informal introduction to game theory with a particular eye toward coordination games and cooperation. I have tried to keep the mathematics as accessible as possible, but a little math is inevitable. Such a brief chapter cannot do the work of a full introduction to game theory, but I hope enough of the ideas are introduced that a general reader, unfamiliar with the theory of games, can benefit from the chapter and comfortably read the rest of the book.

Chapter 5 turns to a particular application of game theory: games as a model of formal logic. The correspondence between first-order logic and the theory of zero-sum games of perfect information is delightful. I give a logic whose "formulas" are English sentences and show how to evaluate them using games. This is only a small corner of the broader relation between games and logic, but I think that general readers will enjoy seeing a small part of this bigger subject. The use of zero-sum games in logic, as opposed to the coordination games in pragmatics, is also useful in understanding how semantics and pragmatics differ from each other.

Part III turns to the development of bounded rationality. Decisions are constantly made under computational bounds; we do not have perfect knowledge, and we must often make leaps of faith. Chapter 6 explores the problem of common knowledge in some detail. Game theory offers a model of common knowledge, since the players are assumed to know the game they are playing. The mutual knowledge that the speaker and hearer must have in order to communicate can be incorporated as part of the game they are playing. We can avoid the puzzle of infinite regress: my model of your knowledge includes your model of my knowledge, which includes my knowledge of your knowledge, and so on. We can assume that the required knowledge is included in the game and is therefore public. Because of inherent bounds on knowledge, the model of the game is always imperfect. We can use these bounds to think about a variety of interesting phenomena, including linguistic accommodation at one end of the spectrum and misunderstandings at the other. In fact, we can use bounds on common knowledge to model some of the pragmatics of definite descriptions.

Chapter 7 turns to games of partial information, a type of game developed by Prashant Parikh. These games are particularly useful in modeling communication, linguistic and otherwise. I use games of partial information to develop a neo-Gricean model of word finding. This model is sensitive to both the absolute frequencies of lexical items and to the contribution of context. The games are used to model the lexical side of garden path sentences. As a further illustration of games of partial information, I develop a small model of irony that involves selecting a high payoff state or avoiding risk. In the former case, the speaker and hearer know enough about each other to get the implication of irony; in the latter case, the irony is missed and only the literal content of the sentence is taken.

Chapter 8 illustrates the use of games with two examples. First, a model of discourse pronouns is developed. This model forms only a part of a broader theory of discourse anaphora; the latter theory is beyond the scope of the present work. However, a game analysis can be developed for some simple texts; doing so allows us to identify principles that can be generalized to the study of anaphora in general. The chapter continues with a discussion of the analysis of politeness and how we can use politeness to elucidate conversational implicature. Once again, I can only allude to a larger theory that goes well beyond the present scope; nevertheless, the example illustrates the game methodology and suggests some avenues for future research.

The last chapter turns to the problem of lexical content given a context. We will use the important notion of focal point, due ultimately to Thomas Schelling, to develop a system of social coordination of reference. I argue that these focal points are conventionalized, via social practice, into the concepts associated with lexical items. Furthermore, the process of conventionalizing these focal points has an economic and ecological character whose logic can be formalized, understood, and tested empirically. The resulting system gives insight into the difference between homophony, when two unrelated meanings are associated with the same phonological sign, and polysemy, where a single form extends its hegemony over a semantic space. The chapter ends with some thoughts on how to simulate such a system; the resulting approach takes meaning to be an emergent property of social signaling.

I have found game theory to be a useful way of thinking about linguistic meaning and have more than once been charmed by avenues it has opened up to me. I hope readers will be similarly charmed. The resulting social theory of meaning is a useful anodyne to the relentlessly solipsistic world we have come to inhabit. I hope readers will come and join the fun.

Acknowledgments

Throughout my career, I have had the great good fortune to be surrounded by interesting people who have constantly challenged my most fondly held opinions. It seems to me that in keeping with one of the themes of this book, real meaning emerges by being in a community. So many people have helped me think about things that I fear I will inevitably forget to thank them all. I hope I will be forgiven if my memory slips and I leave someone out.

The University of Pennsylvania has provided both a fruitful environment for my various projects and the means to pursue them. I want to particularly thank David Embick and Bill Labov for their thoughtful comments on my various ideas. They have both said things that have challenged my thinking; my work has improved because of their help. Tony Kroch has been generous with his time and ideas. Gillian Sankoff has made me think about language change and language learning over a lifetime; her work made me rethink some of my deeply held assumptions.

I have had the great good fortune to work with Murray Grossman on the neuroanatomy of quantification and number sense and, more recently, on social and linguistic decision making. Our joint experiments, as well as his work with patients, have kept me aware that, in the end, all of this is about real people. I also want to thank the people in his lab, particularly Corey McMillan and Katya Rascovsky. Our weekly lab meetings have kept me grounded in the empirical. Thanks are also due to the National Institutes of Health, which funded our research, under grant NS44266, and whose generosity has made it possible for me to pursue some of the work reported here.

I owe a great deal to my students, both graduate and undergraduate. I want to particularly thank Laia Mayol, Ian Ross, and Neville Ryant, who have been actively involved in working on games and language.

Caitlin Light and Jon Stevens have also contributed to my thinking about language, learning, and strategic decision making.

A Weiler Faculty Research Fellowship allowed me to take time off from teaching to actually produce this manuscript. I'm not sure I could have done it without the generosity of the fellowship.

I should also thank Prashant Parikh, both for his thoughtful discussions and his encouragement. He has been an enormous influence on my thinking about the strategic aspects of meaning.

A number of people have read and commented on the manuscript of this book. I want to particularly thank Connie Abrams and Jakub Szymanik for their close reading of the book. Connie reminded me of the dignity of the reader and helped immeasurably with both the style and substance of the book. Jakub read the book closely and made a number of suggestions that have improved the result. Ed Keenan caught some crucial errors as well as sending some useful encouragement.

I also want to give a big thanks to two anonymous reviewers; there is nothing better than being taken seriously, and they took the book seriously. Their reviews were thoughtful and substantive, and I owe them a great deal. I want to thank my editor, Ada Brunstein, for providing me with such reviewers.

My greatest thanks goes to the person to whom I owe the most, Jennifer Hasty. She has been my best critic and biggest fan. Whenever I grew discouraged, she was there to lift me. She is my partner and my inspiration. She has also kept me grounded. I am the most fortunate of men to have her.

Finally, just as I was finishing the manuscript, our lives were enlivened by the arrival of our daughter, Thisbe Genevieve Hasty Clark. Most radiant of children, she is a delight. It gives me great pleasure to think that one day, years from now, she will be able to go to the library, open this book to this very page, and see how much I love her. Be well, my darling.

THE SOCIAL SIDE OF MEANING

1 Platonic Heaven

Sometimes you are drawn to something by pure mystification. The biologist might be baffled by the emergence of life from brute matter. How could it be that a cell could support life? The neuroscientist might be dazzled by the emergence of thought and consciousness from the neurochemistry and topology of the nervous system. How is it that light, hitting the eye, results in the *experience* of red?

The world is surely filled with more than enough to dumbfound and amaze for a lifetime. But the object of wonder that has most enticed me is meaning. What is meaning, and how is it possible for me to mean something? Particularly puzzling is the fact that I can mean something by making noises with my mouth, or making marks on paper, or moving my hands in particular ways. In this chapter, I lay out a way of thinking about meaning in language that has motivated an enormous amount of work in theoretical linguistics. I used to believe fervently in it, but I don't anymore. For reasons I give in chapter 2, I've become a sceptic. In chapter 3, I motivate an economic theory of meaning that lays the foundations for the work I really want to talk about: using the theory of games to think about strategic aspects of meaning.

The Puzzle of Reference

For the moment, though, let's revel in the mystification of meaning. Once, I had to go to a conference in Prague, a city I had heard *spoken of*, had *read about*, but had never seen. I told a travel agent that I had to go to Prague, and after some more noises, I found myself in possession of an airplane ticket. I made a phone call and spoke to some unseen and, to me at least, unknown person who purported to work in a hotel in Prague. I was told, after some negotiating, that I would have a room. All this by moving my tongue and jaw appropriately.

Of course, I wasn't out of the woods yet; perhaps I had accidentally arranged to go to Cleveland. I went to the airport, and there was a plane putatively destined for Prague. Apparently, my discussion with the travel agent had worked; the next thing I knew, I was on the plane in exactly the seat that my travel agent told me she had reserved in my name. The plane took off, and there I was with the presumptive destination of Prague, capital of the Czech Republic.

Once I had landed and cleared customs, I found a cab and *told* the cab driver the address of my hotel. And the honest fellow drove me right to it. There it was exactly as promised. Astonishing! Not only that, but I did indeed have a room there, just as I had negotiated with the clerk by telephone.

Later, with the aid of a map and a guidebook, I confirmed that I was in Prague. There was a river precisely where the map promised the Vltava River would be. I walked across a bridge that purported to be the Charles Bridge and saw Prague Castle up on top of a hill, just where it was supposed to be. Using material in the packet the conference organizers had sent me, I went to an address and found a room where a group of people led me to believe that they were attending precisely the same conference I was supposed to be attending. At the appointed time, I gave a talk. The audience nodded, seeming to understand me. Some even asked questions that were relevant to what I had said; apparently I had communicated something to them. They gave every appearance of grasping my meaning.

There were only two possibilities. One possibility was that somehow, using marks and noises, I had successfully gone to Prague, not Cleveland, and given my talk at the correct conference. A more sinister possibility was that I had fallen victim to an immense conspiracy, some vast prank, and had wandered into some hoax Prague—perhaps Cleveland disguised as Prague—where people pretended to be attending the conference and feigned that they were following my talk.

Rejecting the second possibility as too fantastic—who would benefit from such a conspiracy?—I settled on the reasonable hypothesis that I was in Prague, the capital of the Czech Republic, attending the conference. I had spoken; my meaning had been understood. Somehow I had used noises to solve problems. How can it be?

Use, Mention, and Truth

The reader may well be baffled by my bafflement, but bear with me. There are good autobiographical reasons why I am so puzzled by simple things.

I was born in the southwest of the United States, near the border with Mexico, at a time when *gringos*—mainly white, English-speaking Yankees—were a minority. Most people living there were *chicanos*, although we referred to them as *Mexicans* because they were Hispanic and many of them spoke Spanish. Of course, they had been living in the area since Methuselah was in short pants. We *gringos* were the interlopers, and we were decidedly the linguistic minority; English was a relatively recent transplant to the area, Spanish having been imported there centuries earlier. In reality, my family was living in a colonial situation. The white, English-speaking minority community was economically dominant and largely insulated from the poorer, Hispanic majority. Although my parents were far from wealthy, we could easily afford to have a maid come in from Juárez to do the housekeeping.[1]

A few *chicanos* were able to climb into the middle class; I don't remember my parents socializing with them beyond the requisite low-level civilities. I, however, happily played with the Gonzalez children next door, so much so that my parents, already worried by the pervasiveness of Spanish, grew concerned that I might end up speaking better Spanish than English. They needn't have worried. Everything about the social environment and the local economy at the time pushed me toward learning English. English, after all, was the language of the economically and politically dominant class. For years, I associated Spanish with poverty, the language of the underclass.

As a boy, I was surrounded, or so it seemed to me, by the largely impenetrable code of Spanish. When I was small, my mother and I would venture out of our little Anglophonic island and suddenly be immersed in a completely mysterious world where everyone spoke a language we didn't understand. When I was older and could venture out on my own, I learned to curse in Spanish, but was otherwise oblivious to it.

Of course, we had obligatory Spanish lessons throughout primary school. I managed to learn very little, but I did take special note of facts like the following:

(1) *Perro* means dog.

As it happened, I completely misunderstood the translation rules like the one in (1). Instead, I understood them as

(2) *Perro* means *dog*.

There's a crucial difference between (1) and (2), one that had a big impact on me.

Philosophers and linguists make a distinction between *using* a word and *mentioning* it. In (3) I use the word *dog* to help me refer to some particular dog:

(3) My neighbor's dog barked all night.

In (4), I use *dog* to refer to the word itself:

(4) *Dog* is a word of one syllable.

That is, I mention the word *dog*. Clearly, it makes no sense to suppose that actual dogs are monosyllabic words. I've actually mentioned *dog* several times on this page; the mention of a word shows the curious ability of language to turn on itself and talk (and think) about itself.

This ability of language to refer to itself has been a source of enormous philosophical puzzlement, as (5) shows:

(5) This sentence is false.

The sentence in (5) is true if it's false and false if it's true. This is an example of the famous liar paradox, which is often taken to be a problem of self-reference, that is, the word *this* in (5) refers to the sentence that contains it. But really the problem lies in language's ability to talk about language, so *self-reference* must be taken in a very broad sense. To see this, look at the sentences in (6):

(6) a. The sentence in (6b) is true.
    b. The sentence in (6a) is false.

Neither sentence in (6) refers to itself, but they still have the flavor of the liar paradox in (5). If (6a) is true, then it must be false. (6a) asserts that (6b) is true. But (6b) says that (6a) is false. So if (6a) is true, then (6b) must be true and (6a) must be false. The two sentence consume each other like an ouroboros. The problem is that we're using language to talk about itself.

All this is just to say that my early confusion has a distinguished philosophical pedigree. Let's take a closer look at my problem. I understood the teacher as saying

(7) *Perro* means *dog*.

This means not that *perro* means in Spanish what *dog* means in English (an actual canine). Instead, it means that the Spanish word *perro* denotes the English word *dog*.

The Language of Thought

Laugh, if you will, at my boyhood theory of Spanish, but I note that it is not without precedent, and in many ways it is an instance of a perfectly respectable theory of meaning. What I decided was that speakers of Spanish were internally speaking English and translating from English to Spanish when they spoke and from Spanish to English when they listened.

Of course, I had to solve the problem of why Spanish speakers often didn't understand English. I concluded that although they were thinking in English, these English thought processes were inaccessible to their conscious minds (I must have absorbed some talk of Freud and the unconscious). So here was my theory. Although Spanish speakers thought in English, English was not accessible to them as a means of communication. They therefore had to frame everything in terms of the language they knew, namely, Spanish. The grammar of Spanish, then, must be a translation manual between Spanish and English.

Imagine how gratified I was to learn, many years later, that a famous Enlightenment philosopher had asserted that French was the language of thought. When asked whether the ancient Romans thought in French, he unflinchingly responded that they must have done so, even though French is a descendant of Latin. I admire his confidence.

It might seem peculiar to say that we arrive at a speaker's meaning by translating what she says into some other language. But if you think about it, if you can translate correctly from, say, Spanish to English or French to Latin, you would need a very thorough understanding of Spanish or French and English or Latin. Furthermore, if I happen to speak English or Latin, then your translation is very informative. The idea of producing a *translation manual*, a manual for translating from a language you don't know to a language you do know, as a kind of theory of meaning has been advocated by the philosopher W.V.O. Quine, for example. In one form or another, the idea has been a very important one for linguists working on meaning, although I think that Quine would probably disagree with the form much of this work has taken.

Now, where I went wrong, some would say, is not in the idea that speakers of Spanish were translating back and forth between some internal mental language and Spanish, but in supposing that English speakers weren't doing so. What if they were translating from their external language, a language that they would have to learn, into a special internal language, a language that they understood from birth and thus didn't need to learn. It might be that there is a kind of internal mental language—a Language of Thought (LOT), or Mentalese, as it is sometimes known—and that when people speak they take a sentence in Mentalese and translate it into whatever language they use to communicate. When they hear a sentence in their external language, they analyze it and translate it into Mentalese.

Of course, no one would necessarily have any direct perception of Mentalese. I have a strong intuition that I think in English, but perhaps I'm aware of my thoughts only after they've been translated from Mentalese to English. All sorts of things go on below the level of conscious awareness. For example, I have no reliable intuitions about how I process visual information. I was surprised to learn that the brain has two different visual systems; one system recognizes where objects are in space, and the other system recognizes what the objects are. The two systems can be independently impaired. Someone with an impairment in the "where system" can recognize an object but can't reliably reach for it; someone with an impairment in the "what system" can reach for the object accurately but can't recognize what it is, even though he may know a lot about the kind of thing the object is.

My point is that brute intuitions about plausibility are not the most reliable way to judge an idea. Instead, we need to think about the empirical consequences of the idea. If the theory fails empirically, then we need to cast about for a better theory. Equally, if there's no way to test the theory—no evidence that could possibly count against the theory—then the theory needs to be rejected. Linguists are in the business of producing theories that can be tested empirically.

So think about the following idea. In understanding a sentence, we translate that sentence into the Language of Thought, and when we want to communicate an idea we translate from the Language of Thought into whatever spoken language we use. Assume that part of our linguistic ability includes rules like the following, a *truth predicate*:

(8) S is true $\Leftrightarrow \Sigma$.

The S in (8) would be a sentence in some natural language like English and the Σ would be a sentence in Mentalese. The double arrow $\Leftrightarrow$ means 'a systematic mapping between', so when I encounter the sentence S, I can apply the procedure indicated by $\Leftrightarrow$ and get the resulting expression Σ in Mentalese.

The basic idea is that the way to work out a theory of meaning for a language like English is to show how to translate from English to Mentalese. Since we understand Mentalese perfectly, the translation would be from an external spoken language into a language we understand. We could then rely on Mentalese, the true and only Language of Thought, to imbue English with meaning.

Readers may think that I merely make things more complicated by adding some mysterious new language to the mix. How can an unseen Language of Thought be empirically tested; isn't there a risk of this being a nontheory with no real empirical import? And what is the word *true* in (8) doing there?

Let's start with the word *true*. Saying what *true* actually means is so difficult as to be well beyond my abilities, but we can rely on the basic intuition that part of knowing the meaning of a sentence is knowing what the world would be like if the sentence were true. We could give a mathematical theory of truth, a theory that lays out how a sentence (or expression in Mentalese) could be true about the world. This might not work as a metaphysical definition of "the true" (whatever that is), but it could say something about the relation between language and the world and, in consequence, how language can carry information about the world.

Suppose I tell you something like

(9) I have a cocker spaniel named Sami.

You have several pieces of information from my utterance. Among other things, you know that I have something called a cocker spaniel and that this particular cocker spaniel answers to the name *Sami*. Of course, you can bring other information to bear on my statement; for example, you might also know that cocker spaniels are a kind of dog.

Now, what I said—that I have a cocker spaniel named Sami—is combined with what you already know—that cocker spaniels are a kind of dog—to *entail* that Sami is a dog and that I have a dog. This notion of *entailment* is defined in terms of truth and is important for understanding things like inference, the ability to combine bits of information to get new bits of information. Entailment can be defined as follows; I've simplified the definition somewhat, but it should be serviceable for now.

(10) **Entailment**
A set of sentences $\{S_0, \ldots, S_i\}$ entails another sentence S_m if and only if sentence S_m must be true whenever all the sentences in $\{S_0, \ldots, S_i\}$ are simultaneously true.

Famous examples of entailment are Aristotelian syllogisms like

All men are mortal.
Socrates is a man.
―――――――――――――
Therefore, Socrates is mortal.

While the syllogisms may seem remote from experience, in fact entailment is used all the time in reasoning about the world. So the notion of truth in (8) is actually an important element in understanding how we use meaning in everyday language. Here we can begin to see a foundational fact for any theory of meaning: we use meanings to do things. Meanings are not simply an assemblage of facts; instead, they are tools for organizing behavior and thought; they are tools for operating on the physical and social world.

Concepts, Mentalese, and the Informational Universe

The next question that comes up regarding (8) is why Σ is couched in terms of Mentalese. Mentalese is a language of concepts. We all live in a physical world buzzing with clouds of particles, radiation of various sorts, and the interplay of fundamental forces. But that isn't the world of our experience. When I look around right now, I see my computer, my desk, a bunch of books, my telephone, my coffee cup, and so on. But these objects are informational things, not fundamental categories of the physical universe. Although a chair is a physical object, its role as a chair involves information; it requires that we recognize its function as a chair.

Where do these informational categories come from? Where does the informational universe come from? And given that we have these informational things, how would they be used in the real world?

My coffee cup must have certain physical properties to work as a coffee cup, but whether it's a coffee cup or not is largely up to me. I could use it as a paperweight, or as a shaving mug, or as a hat, or as a Christmas tree ornament, or as a collar ornament for my dog Sami. The role of my coffee cup in the world is only partly a matter of its physical properties. It has to work as a coffee cup for containing hot liquid, but its role is largely determined by how I fit its use into a broader scheme of things; it's really only a coffee cup if I decide to use it as such. I am the captain of my coffee cup. My mind makes it what it is. (I don't really believe this as stated, but let's go with it for the moment and work out what I could possibly be thinking.) In fact, I might decide that just about any receptacle for liquids could act as a coffee cup. After all, don't billionaires drink champagne from ladies' slippers? What's to stop me from dubbing my shoe a coffee cup and drinking from it?

"Well," you might say, "that may hold for coffee cups and other things that people make. They're artifacts, and their use is a matter of human agency. But what about objects in the natural world, things that aren't artifacts." So let's take the case of a biological category, like "tiger." Is there some obvious physical property of tigers that make them, and nothing else, tigers?[2]

Tigers are striped quadrupeds that engage in predatory behavior; they have whiskers and big sharp teeth and claws. I might add that they are felines, but that category only makes sense inside a theory of biology, so let's set it aside for the moment.

None of the physical properties I mentioned are actually necessary criteria for tigerness. Suppose I had a tiger, Claude. He's a big striped quadruped, and he spends his time hunting. He indeed has big teeth and claws, very sharp and dangerous. So he's a tiger, as described.

But now suppose I take Claude to a laser hair removal center and have all his fur and whiskers lasered off. Claude's bald now, so he's no longer striped but he's still a tiger, near as anyone can tell.

What if I take Claude to a physical therapist who teaches him to walk (all the time) on his hind legs. He's still a tiger, even if he's no longer a quadruped.

Now suppose Claude has a spiritual conversion: predation and meat eating are wrong. From now on, Claude renounces meat and decides to eat grass. To accomplish this, he has his sharp teeth removed and special flat dentures installed, so that he can better grind up the grass with his new teeth. Furthermore, to ensure his pacific ways, he has his claws removed (perhaps I should call him de-Claude). Is he still a tiger? Yes, although at this point he looks and acts nothing like a tiger.

Perhaps Claude is a tiger because he has tiger DNA. But certainly the concept of "tiger" doesn't rely on DNA; after all, the concept of "tiger" was around before anyone had heard of DNA. People are mostly essentialists, I think. Once Claude has been fit into the concept of "tiger," he's treated essentially as a tiger, no matter what his appearance is. It's hard to get him out of that concept unless we open poor Claude up and discover that he was, all along, a robot. Then he becomes a "robot tiger," I suppose.[3]

Although it no doubt has some support from the physical world, perhaps "tiger" is largely an informational category. That category is as much about how we think about the world as it is about the physical properties of the world; it seems that we have found the Language of Thought made manifest in the world.

Another example of the interaction of mind and world is our sense of number. Right now, I have four books on my desk, next to my computer. There is some evidence that I perceive the number four as an independent category; that is, part of my brain is devoted to the direct perception of number. Numbers may exist independently of our minds, or they may be constructed by us, but there can be no doubt that there is a specific neurobiological structure devoted to the perception of number. It's hard to see any physical world constant that would correlate with fourness. Nevertheless, we are able to extract numbers from the environment when called upon to do so. It seems as though number sense is a conceptual system that exists by virtue of the structure of our brains.

Our day-to-day talk is larded with all sorts of informational categories that have a very tenuous relation to the physical world. Suppose, watching the stock market fluctuate wildly in light of the crash in credit markets, I utter the following:

(11) The proposition that the invisible hand of the free market converts individual greed into social good is fundamentally flawed.

Surely, no one would expect what I said in (11) to be transparently supported by the physical world. It is riddled with concepts like "the invisible hand," "greed," and "social good," none of which have any transparent relation to physics.

This is really a very old point. In the first half of the twentieth century, a group of philosophers, the logical positivists, thought they could replace loose talk expressed in terms of abstract categories with precise talk grounded in physical measurement. The movement was short-lived, although quite influential; it didn't take long to realize that abstract categories are indispensable to our understanding of the world.

Language and the World

Of course, we're not free to treat concepts in any way we please. Concepts have to be tied to the world somehow. Example (8) showed a translation of a sentence S of an external language into a sentence Σ of Mentalese using a truth statement:

(12) S is true $\Leftrightarrow \Sigma$.

But this translation must be supported by another translation,

(13) Σ is true ⇔ TC,

where TC is a specification of the conditions under which the Mentalese sentence Σ holds true. That is, Mentalese, if it is to be useful at all, must be what is called an interpreted language that connects to the world. It needs to be interpreted because we cannot take its terms and predicates as basic. If we did take Mentalese as basic, a primitive, then our spoken language—the language being interpreted by Mentalese—would be unable to convey information about the world. But the fact that I made it to Prague, not Cleveland, shows that my language does have a connection with the world; if I'm to operate in the world and use language to learn about it and negotiate it, there must be a connection to the world.

In short, although language may translate to concepts, these concepts must relate to the world. We know this because we're able to coordinate our actions in the world using language. This means that Mentalese should be interpreted relative to a world model that is considered external to the speakers of a language:

Language → Mentalese → World model.

The world model would concern more than just the physical world; it would include abstract things like number and time, for example. But, crucially, it wouldn't be an internal, private representation of the world. It would be a shared public space, available to all speakers, that could be used to coordinate their verbal and conceptual behaviors. That way, when I say "dog" or "coffee cup" or "Prague," the corresponding concepts in Mentalese—DOG or COFFEE CUP or PRAGUE—would pick out dogs and coffee cups and Prague in the world model.

Platonic Heaven in a Box

Now, you might object that I just added more work. English must now be interpreted relative to Mentalese, and Mentalese relative to some model of the world. The following would doubtless be easier:

Language → World model.

Just skip the middleman and go directly to the world. I have some sympathy for this position, but let me try to give an answer that's fair to the Mentalese theorist.

We need a theory of linguistic meaning that properly connects language to both human reasoning and human action. Mentalese would be a common cognitive language that could connect these disparate areas and organize them relative to the world. Furthermore, our concepts could be part of the world model, providing a way of making our private thoughts and opinions public. All this would be much harder to do without the common, mind-internal language of Mentalese.

Mentalese, of course, can't be exactly like a natural language. It's a language of mental representation that everyone uses but no one speaks. To make Mentalese work, we all need to have the same Mentalese concepts and agree as to how these Mentalese concepts pick out things in the world model, the simulacrum of the real world. This is a pretty tall order.

We can get some handle on the problem by consulting Plato's dialogue *Cratylus*. Hermogenes accosts Socrates and asks his help in solving a problem. Cratylus, the teacher of Hermogenes, teaches that there is a right and wrong way to call things. That is, each thing has a unique correct description, according to Cratylus. Protagoras, another teacher, claims that "man is the measure of all things." That is, there is no unique right or wrong way to call things: I use *dog* and the French use *chien*, and that's just the way it is. Neither of us is uniquely right; we're both right. Hermogenes wants Socrates to declare who is right: Cratylus or Protagoras.

The argument between Cratylus and Protagoras is really about whether linguistic signs are conventional (Protagoras's position) or natural (Cratylus's position, with which Socrates agrees). Note that whatever the signs of Mentalese are, they can't be conventional. Conventional things are arrived at through public practice, and there is nothing public about the signs of Mentalese; they're entirely internal to the brain or mind. We can only see Mentalese signs indirectly by virtue of our language use.

Early in the dialogue, Socrates lays the groundwork for his case that signs are natural:

Socrates But how about truth, then? You would acknowledge that there is in words a true and a false?
Hermogenes Certainly.
Socrates And there are true and false propositions?
Hermogenes To be sure.
Socrates And a true proposition says that which is, and a false proposition says that which is not?
Hermogenes Yes, what other answer is possible?
Socrates Then in a proposition there is a true and false?
Hermogenes Certainly.

Socrates But is a proposition true as a whole only, and are the parts untrue?
Hermogenes No, the parts are true as well as the whole.
Socrates Would you say the large parts and not the smaller ones, or every part?
Hermogenes I should say that every part is true.
Socrates Is a proposition resolvable into any part smaller than a name?
Hermogenes No, that is the smallest.
Socrates Then the name is a part of the true proposition?
Hermogenes Yes.
Socrates Yes, and a true part, as you say.
Hermogenes Yes.
Socrates And is not the part of a falsehood also a falsehood?
Hermogenes Yes.
Socrates Then, if propositions may be true and false, names may be true and false?
Hermogenes So we must infer.

In other words, a true sentence will be true in virtue of the truth of each and every one of its constituent parts. This passage anticipates an important idea in linguistics and the philosophy of language:

(14) ***Compositionality***
   The meaning of a phrase is a function of the meanings of its parts and their mode of combination.

This is an extremely plausible idea that accounts for how each of us is capable of understanding new sentences. According to compositionality, I need to know the meanings of the atomic parts of the sentence, say, individual words, and I need to know how they combine to make up the whole sentence. That is, if I know what the words mean and I have a grammar that tells me how to combine words into sentences, then I can work out the meanings of sentences.

It is clear where Socrates is going with this argument. If a sentence is true, it must be because its parts are true. If the parts are true, it must be because their parts are true. And so on, down to the atomic level of words. It must be, then, that words are true of the objects they denote.

According to Socrates, there is a right and proper name for each thing, such name given by an "artificer of names" or "legislator" who skillfully associates with each thing the name it should have by nature:

Socrates Then, Hermogenes, I should say that the giving of names can be no such light matter as you fancy, or the work of light or chance persons. And Cratylus is right in saying that things have names by nature, and that not every man is an artificer of names, but he only who looks to the name which each thing by nature has, and is able to express the true forms of things in letters and syllables.

There follows a lot of fanciful Greek etymology, designed to get at the true nature of things.

I doubt that many people would defend the natural theory of names that Socrates and Plato advance. It goes well with the idea of a Platonic heaven, where true forms dwell. Certainly, few would want to say that Greek or French or English words are more natural than those in another language. Everyone agrees that words are arbitrary symbols.

But what about the symbols of Mentalese? Mentalese is not supposed to vary in the way that natural languages vary. Everyone must be equipped with the same Mentalese.

A Mentalese theoretician would, I think, have to agree with Socrates that the signs of Mentalese are natural, not conventional. He would argue, I think, that the "artificer of names" is none other than evolution. Evolutionary psychology, which seeks to explain aspects of mind in terms of evolutionary theory, holds that we've evolved to have certain organs of perception, to act in certain ways in the world, and to think of the world in particular ways. Presumably, the way we think, perceive, and act has been of benefit to our species, aiding survival and reproduction. Hominid A, equipped with proto-Mentalese, is able to categorize and conceptualize the world in a useful way. She is better able to reason from the information she perceives. This adds to her reproductive success so that she passes on proto-Mentalese to her offspring. Hominid B is a clod with no internal representational capacity. He can't efficiently categorize or reason about the world. Being an ignorant oaf, he lacks hominid A's survival edge and is doomed.

Eventually, hominid A's proto-Mentalese would be passed on and modified into Mentalese. Mentalese itself, if it exists, would have to be part of our biological endowment. In other words, each of us would have to be born with an innate representational system that underlies our reasoning and action, the Language of Thought.

Inferences and Mentalese

I have some doubts about whether Mentalese predicates are heritable traits, but let's take a concrete example. Everybody has the concept of causation as part of their internal representational system. Suppose there's a Mentalese expression, CAUSE, that we're all born with. It would work as follows: an AGENT would CAUSE an EVENT to transpire. In Mentalese,

(15) (CAUSE(EVENT))(AGENT).

Equally, we all have the notion, as part of our innate endowment, that things die, so Mentalese would include DIE. The thing that dies is not an AGENT; call it a THEME. The Mentalese expression would be

(16) DIE(THEME).

Now, we would also know

(17) DIE is a kind of EVENT.

We would learn that the English word *kill* means that the AGENT of *kill* caused the PATIENT to die. Thus,

(18) AGENT kill PATIENT ⇔ (CAUSE(DIE(PATIENT)))(AGENT).

Putting all this together, when a speaker of English hears

(19) John killed Bill.

she would translate it to the Mentalese expression

(20) (CAUSE(DIE(BILL)))(JOHN),

where JOHN is the Mentalese symbol for John, and BILL is the Mentalese symbol for Bill. Because Mentalese is interpreted relative to a world model, she would know that John caused Bill to die in the world.

Even better, as an innately endowed speaker of Mentalese and a competent speaker of English, she might have access to the following rule:

(21) If (CAUSE(EVENT))(AGENT) then EVENT is true.

The rule in (21) is called a *meaning postulate*. It places a constraint on how causation is interpreted; if an event is caused, then that event actually has to happen.

Thus, we have the following translation from English to Mentalese:

(22) "John killed Bill" is true ⇔ (CAUSE(DIE(BILL)))(JOHN).

We also have the following correspondence from Mentalese to the world model:

(23) "(CAUSE(DIE(BILL)))(JOHN)" is true ⇔ John actually caused Bill to die in the world model.

Armed with the meaning postulate in (21), we can conclude that if John killed Bill, then Bill is dead. But this is an example of entailment. So this system of translations and meaning postulates actually can support an account of how we might reason with language.

When I was a boy, I had settled on the idea of English as Mentalese. It seemed utterly natural to me that *dog* meant dog and regrettable that Spanish speakers had to translate dog to *perro*.

Still, sometimes I would lie out on the grass in the backyard, watch the clouds, and repeat to myself "dog...dog...dog..." until the word itself disintegrated into just so much sonic nonsense. Then, the connection between *dog* and dog became mysterious, something to be wondered at. Why, I wondered, would *dog* mean dog?

And therein lies a problem. Suppose I could have repeated the Mentalese predicate DOG to myself. Is its connection to an actual dog any sturdier than the connection between *dog* and dog? The great artificer of names seems powerless here; how did I connect my mind-internal concept of DOG with that dog out in the real world?

Further Reading

A good place to start reading about truth is Blackburn's *Truth: A Guide* (2005). The translation theory of meaning is discussed in Quine's *Word and Object* (1960), and is critiqued in an article by Davidson (1974). The liar sentence in (5) is well-known; the multiple-sentence liar in (6) is adapted from Gupta and Belnap (1993). The true master of the liar paradox is Raymond Smullyan. His puzzle books are an encyclopedia of self-reference, but his masterwork is Smullyan (2009), which provides a kind of logical cosmology of lying and truth telling

Jerry Fodor has been an articulate champion of the Language of Thought; see Fodor (1975). I remember going to hear him as an undergraduate and being impressed when in response to a question from the audience, he argued that Neanderthals had the concept of "carburetor" as part of their innate Language of Thought. It's worth reading Fodor in tandem with Cowie's (1999) book, which gives a balanced discussion of nativism.

The translation statement in (8) is a deliberate conflation of an idea from Tarski (1983), who gave a mathematical definition of truth in formalized languages like logic. The idea is to transfer Tarski's approach to the Language of Thought.

A good discussion of number sense can be found in Dehaene (1997). Murray Grossman, a neurologist at the University of Pennsylvania, and I have worried about the relation between language and number sense; see Clark and Grossman (2007) for an interim report on the neurobiological underpinnings of language and number.

A good discussion of logical positivism and its downfall can be found in Soames (2003). Ray Jackendoff and Steven Pinker are both ardent defenders of Mentalese within linguistics. Fodor famously wrote a paper called "Three Reasons for Not Deriving 'Kill' from 'Cause to Die'" (1970), so he would surely not endorse my Mentalese analysis of *kill*. I certainly don't want to tar him with the brush of lexical decomposition (his theory is much more subtle). Nevertheless, the particular decompositional theory of meaning I described has wide currency in linguistics. For a very sophisticated version, see Hale and Keyser (2002) and the references cited there. See Jackendoff (1983) for a clear statement of Jackendoff's views. Pinker (1994) provides a widely read, very accessible discussion of generative grammar along with Mentalese. His more recent (2007) book delves into Mentalese and the structure of the lexicon.

Compositionality is often attributed to the nineteenth-century logician Gottlob Frege, although he didn't spell out exactly what he meant. See Dummett (1981) for some discussion.

A thorough discussion of inferencing and entailment can be found in any good introduction to logic. I'm particularly fond of the introductory text by Barwise and Etchemendy (1989).

2 My Fall from Platonic Heaven

The theory outlined in chapter 1, the Mentalese theory, is a formidable one. Its intellectual roots run deep. One sees it anticipated in Plato and Kant. It has absorbed ideas from the philosophy of mathematics and logic, computation theory, and artificial intelligence. No one should take it lightly; without this theory, linguistics as we know it today would look radically different.

My formulation may be a bit oversimplified, but I think it's fair to say that many linguists believe some version of it. I fervently believed it as a graduate student and defended it and taught it when I became a faculty member.

The theory takes the mind (or brain) as a computational device. What exactly does that mean? At the very least, a computational device is a system that has a set of symbols and operations defined to manipulate those symbols. In chapter 1, I imagined that the human capacity for language was one kind of computational system. It would have an internal vocabulary that could be used to specify a grammar, namely, a set of rules that would tell the system how to construct and parse sentences.

Phrase Structure Grammar

Let's take a simple example of a grammar and work out the relation between the rules of grammar and their meaning in Mentalese. Figure 2.1 shows a very simple grammar—called a *context-free phrase structure grammar*—for a few sentences of English. Each line in the figure is a single rule. The arrow symbol, →, is either an instruction to replace the symbol on the left-hand side of the arrow with the string on the right-hand side, or an instruction that allows the symbol on the left-hand side of the arrow to be replaced by a single choice from the options listed between the curly braces, { and }. The symbols on either side of the arrow

```
S → NP VP
NP → Det Noun
VP → V_Intrans
VP → V_Trans NP
Det → {the, a, every, some, no, all}
Noun → {tiger, monkey, human}
NP → {Alice, Bill, John, Mary}
V_Intrans → {slept, walked, snored}
V_Trans → {saw, licked, ate, killed}
```

Figure 2.1
A Very Simple Grammar

are the symbols of the computational system, and the operation is specified by the arrow; it is either the concatenation—the stringing together—of symbols or the choice of a single symbol from a set of possibilities.

In order to construct a sentence from this grammar, we start with the symbol S (for sentence):

S

The system says that we can replace the S symbol with the string "NP VP" (for noun phrase and verb phrase):

NP VP

The rules in figure 2.1 allow us to replace NP with the string "Det Noun" (Det indicates determiner; see figure 2.1 for examples):

Det Noun VP

We are allowed to replace Det by *the*:

the Noun VP

and replace Noun by *monkey*:

the monkey VP

VP can be replaced by $V_{Intrans}$, where *Intrans* is short for *intransitive* and means there is no object of the action named in the verb:

the monkey $V_{Intrans}$

Finally, $V_{Intrans}$ can be replaced by *snored* to yield

the monkey snored

which, while not exactly Shakespeare, still counts as a grammatical sentence of English.

Usually, linguists prefer to show the derivation (or parse) of a sentence in terms of a tree, which is neutral between building the sentence and assigning the sentence a parse. The root of the tree is the symbol we started with, S, and under each symbol is the string that replaces the symbol. The root of the tree is at the top and the tree grows down. So the tree for *the monkey snored* is the following:

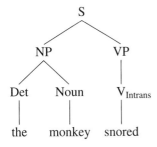

At every level the tree corresponds to steps in the construction of the sentence by the computational system—the grammar—that was specified in figure 2.1.

Of course, a more adequate grammar would be much more complex, but the simple grammar suffices to make a few points. Recalling the example for *kill* that ended chapter 1, readers can verify that the simple grammar in figure 2.1 allows the system to build the tree in (1):

(1)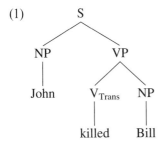

Grammar and Compositionality

The central idea of chapter 1 was that sentences of English can be translated into expressions of Mentalese (assume that the computational system of the mind/brain knows how to handle them). The crucial step was to suppose that the system was compositional in the sense of (14) in chapter 1, repeated here:

(2) **Compositionality**
The meaning of a phrase is a function of the meanings of its parts and their mode of combination.

So first we specify what the atomic parts of the grammatical system mean in Mentalese and then how they combine. Suppose the atomic parts of the grammar are words. The first thing we need to specify is what the words translate to in Mentalese.

Suppose the names *John* and *Bill* denote Mentalese names:

(3) John ⇒ JOHN.
    Bill ⇒ BILL.

Specifying the meaning of *killed* is a bit trickier. Ignoring the tense for the time being, I'll stipulate the meaning, then explain later:

(4) killed ⇒ $\hat{x}\hat{y}(\text{CAUSE}(\text{DIE})(x))(y)$.

For the moment, read the symbol $\hat{x}$ as meaning 'I'm looking for something to replace the x in the following string'.

Now I can say what "mode of combination" means in (2). When two phrases concatenate, as when *kill* concatenates with *Bill*, the function named by *kill* applies to the thing named by *Bill*, in other words,

(5) kill Bill ⇒ $\hat{y}(\text{CAUSE}(\text{DIE})(\text{BILL}))(y)$.

Notice that I replaced the x in the translation of *killed* in (4) with the translation of *Bill*, just as instructed by the $\hat{x}$ symbol. This can be done one more time to get the translation of the whole sentence:

(6) John killed Bill ⇒ (CAUSE(DIE)(BILL))(JOHN).

The whole process is somewhat clearer if it is shown as a tree:

(7)

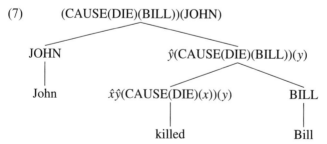

At the root of the tree in (7) is the translation of the whole sentence into Mentalese. The leaves—the elements the farthest away from the root— are the actual lexical items of English. So, reading the leaf nodes from left to right in (7), we get the English sentence *John killed Bill*. Directly

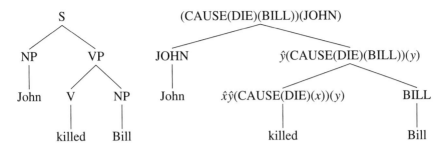

Figure 2.2
Syntactic Parse and Logical Form Comparison

above each English word is its translation into Mentalese. *Bill* is translated to BILL and *killed* is translated to $\hat{x}\hat{y}(\text{CAUSE}(\text{DIE})(x))(y)$—this last is a relation between y things that cause x things to die.

Right above the Mentalese translations BILL and $\hat{x}\hat{y}(\text{CAUSE}(\text{DIE})(x))(y)$ is their composition. The syntactic operation of concatenating a verb and its object corresponds to the semantic operation of applying the function named by the verb to the entity named by the object. In this case, we get $\hat{y}(\text{CAUSE}(\text{DIE})(\text{BILL}))(y)$. Semantically, this new item is a function that designates the set of things that killed Bill.

Next, the function $\hat{y}(\text{CAUSE}(\text{DIE})(\text{BILL}))(y)$ is combined with JOHN, the Mentalese translation of *John*. Syntactically, the predicate phrase *killed Bill* is combined with the subject *John*. Semantically, the function $\hat{y}(\text{CAUSE}(\text{DIE})(\text{BILL}))(y)$ applies to JOHN. If JOHN is in the set of things that killed Bill, then the function returns "true" and one can say that the proposition named by (CAUSE(DIE)(BILL))(JOHN) is true.

This example illustrates the basic properties of one of the most influential theories of meaning in linguistics. I'll call the tree in (7) a *logical form* (LF). Many people would disagree with aspects of the theory I've presented here, but the essential elements are that every syntactic operation has a corresponding semantic operation. This can be seen if the syntactic analysis of *John killed Bill* is compared to the LF (figure 2.2).

Thinking and Computing

The side-by-side comparison of the trees in figure 2.2 makes the point very directly. Every grammatical move has a corresponding effect on meaning. Each operation in the grammar makes a contribution to the interpretation of the sentence. Further, every point in the syntactic parse tree of a sentence has a corresponding point in the logical form of the sentence. This illustrates the genius of compositionality. If I know what the words mean, and I know how the words are put together, then I know what the sentence means. Of course, working out all the details of this connection between grammar and meaning takes a lot of careful, detailed work, but the main idea is simple enough.

For example, having gotten to the proposition associated with *John killed Bill*, we still need to state some rules about meanings. For example, we need to formulate meaning postulates (see chapter 1). These are general rules regarding the connections between meanings. So we have the following:

(8) If (CAUSE(EVENT))(AGENT) then EVENT is true.

In order to make this work, some other things are needed. For example,

(9) DIE(x) is an instance of an EVENT.

As shown in chapter 1, these meaning postulates would allow us to conclude, from "John killed Bill" to John caused Bill to die; from "John caused Bill to die" to Bill died. That is, the meaning postulates would support a whole system of inference that would allow us to use language to gain information about the world.

Although there is a lot of detailed work to do, the basic theory is relatively simple. We need just three kinds of things:

• Translation rules that map from words in a natural language like English to symbols in a system of mental representations, Mentalese.
• Composition rules that specify how to compose expressions of Mentalese into new expressions of Mentalese; these rules are of the same structure as the rules of grammar in accordance with the principle of compositionality.
• Meaning postulates that connect these symbols into an inferential network.

The whole network of rules forms a computational system. We know what the basic symbols are and how to perform operations on these symbols. The whole computational system is subject to a very powerful computer metaphor. Fodor (1980) gives a cogent summary of this view:

Insofar as we think of mental processes as computational (hence as formal operations defined on representations) it will be natural to take the mind to be, inter alia, a kind of computer. That is, we will think of the mind as carrying out whatever symbol manipulations are constitutive of the hypothesized computational processes. To a first approximation, we may thus construe mental operations as pretty directly analogous to those of a Turing Machine. There is, for example, a working memory (corresponding to a tape) and there are capacities for scanning and altering the contents of the memory (corresponding to the operations of reading and writing on the tape). If we want to extend the computational metaphor by providing access to information about the environment, we can think of the computer as having access to "oracles" which serve, on occasion, to enter information in the memory. On the intended interpretation of this model, these oracles are analogues to the senses. In particular, they are assumed to be transducers, in that what they write on the tape is determined solely by the ambient environment energies that impinge on them.

In other words, we can think of the mind/brain as a kind of computer. The brain might correspond to the hardware; it is capable of performing certain operations like writing things to memory, or retrieving them, or altering them in some simple way. The mind would be comparable to software. It would organize the simple operations into actual thought, just as grammar organizes language into useful information.

The whole computational theory of mind is incredibly seductive. It promises a kind of Newtonian mechanics of the mind. Given a mental representation, we can define operations on it that would give us—as sure as night follows day—another mental representation. Our mental lives would be a series of lawful steps from representation to representation. We have the promise that we can disassemble mental activity into its component operations; ultimately, we should be able to perfectly simulate these operations in a physical system like a computer.

The real genius here is that it transposes the world into the mind. Our reasoning about the world would be operations not on the world itself but on our internal Mentalese representations of the world. Pay particular attention to what Fodor says about transducers. When I look at a flower, I don't have direct experience of the flower. Instead, the light reflected by the flower hits my eyes and is transduced via my retinas, optic nerves, and various dedicated visual processing areas in the back of my brain, resulting ultimately in my experience of the flower. This experience—the qualia of the flower—is a mental object, not the flower itself. We do not experience the world as it is; there is no such thing as a simple, direct experience. Instead, we experience the world transduced through our senses and reconstructed in our heads as mental representations. The flower itself might as well be on another planet. This is the *formality condition*, which says, roughly, that the symbols we use—not their content in the real world—are all the computational system needs. The whole magnificent universe is forever separated from us—the real us that is our minds—available only via transducers, our senses.

The Heaven in Your Head

It is as though, in playing a game of chess, the pieces corresponded to actual armies out in the world. My movement of the pieces would be like the movements of the armies, the board itself a schematic map of the world. My game would become a simulation of the world itself.

But then the world could melt away into nothing, leaving me with just the game. What difference would it make? Everything I need to know about the world is contained in the game alone. The external world can vanish, at this point, and I would still generate the same behaviors and internal states on the basis of the symbols and the computational system. Fodor (1980) calls this *methodological solipsism*:

> I'm saying, in effect, that the formality condition, viewed in this context, is tantamount to a sort of methodological solipsism. If mental processes are formal, they have access only to the formal properties of such representations of the environment as the senses provide. Hence, they have no access to the *semantic* properties of such representations, including the property of being true, of having referents or, indeed, the property of being representations *of the environment*.

Once you recover from the shock of the prospect of such absolute isolation, the position is perfectly internally consistent. Our experience of the world is always indirect. The simplest visual impression that we experience is far from direct; rather, it is the outcome of a long series of computational steps. We have no direct, computationally unmediated experience of anything. We might, in fact, be brains in vats, our experience fed to us by mad scientists. There is nothing contradictory about this. In fact, methodological solipsism is maddeningly consistent. The question is whether it is the best way of thinking about our linguistic knowledge and behavior. I suspect that methodological solipsism, although it has proven useful, will ultimately wind up a dead end. The rest of this chapter discusses this problem, but chapter 3 explores evidence that meaning is largely determined by factors that are external to the mind, in contradiction to methodological solipsism.

The computer metaphor is strangely reminiscent of Plato's Allegory of the Cave in *The Republic*. In this allegory, we are asked to imagine a group of prisoners chained in a cave in such a way that they can only see the wall of the cave in front of them. They have been kept this way since birth, so they have no experience of anything outside the cave. Behind them, where they can't look, is a large fire that casts light and shadow on the wall, and between the prisoners and the fire is a raised walkway along which move puppets of people and things. All the prisoners can see, though, is the shadows of the puppets on the wall.

A prisoner is released from bondage and allowed to look at the world outside the cave. At first, the prisoner is confused, baffled by the new sensory experiences. But eventually he adapts and sees things as they are rather than as shadows on the wall of the cave. Plato wrote,

> The prison dwelling corresponds to the region revealed to us through the sense of sight, and the fire-light within it to the power of the Sun. The ascent to see things in the upper world you may take as standing for the upward journey of the soul into the region of the intelligible; then you will be in possession of what I surmise, since that is what you wish to be told. Heaven knows whether it is true; but this, at any rate, is how it appears to me. In the world of knowledge, the last thing to be perceived and only with great difficulty is the essential Form of Goodness. Once it is perceived, the conclusion must follow that, for all things, this is the cause of whatever is right and good; in the visible world it gives birth to light and to the lord of light, while it is itself sovereign in the intelligible world and the parent of intelligence and truth. Without having had a vision of this Form no one can act with wisdom, either in his own life or in matters of state.

In order to have true knowledge, then, we must have access to the universe of Forms—Platonic heaven. What we see, the evidence provided by our transducers, is inadequate unless we can relate it to the pure universe of Forms that underlies it.

Now, our computational mind/brain is like the prisoners in Plato's allegory. The prisoners are like central processing units, equipped with working memory and a program that tells them how to work with the images they see on the wall (their input). The wall of the cave and the fire correspond to the transducers that map sensory data to mental representations. The input data would be the march of puppets along the raised walkway. What about the world of Forms? I can only imagine that the pure Forms of the allegory are the concepts and categories provided by Mentalese.

In Plato's allegory, Form is external to the individual; Form is a realm of perfection—the perfect point, line, triangle, or sphere. What we experience is only a corrupt approximation of these perfect Forms. None of us ever actually experiences a true point or a true sphere, yet we can relate what we do experience to these Forms. How would we know about these Forms, though? Our experience, after all, would be of the corrupted examples that surround us; how would we go from imperfect real-world things to their perfect correspondents? Well, according to Plato, the soul—the freed prisoner in his allegory—would travel to the realm of perfect Form. Recognizing that a real-world ball is related to the pure Form sphere would be a matter of realization, of remembering some aspect of the heaven of perfect Form.

What Mentalese gives us is a portable heaven of perfect Form between our ears. Each of us has access to pure Forms by virtue of having Mentalese, the ability to categorize experience according to our mental representations of the concepts that underlie our cognition. Of course, our ability to do this is thanks to the great artificer of names, our biological endowment. Just as learning the pure Forms is both impossible and unnecessary, so learning the pure concepts that make up Mentalese would be impossible and unnecessary.

Think of it this way. A human infant is not a *tabula rasa*, a blank slate upon which experience can imprint anything. If that were true, the infant would have to commence learning from scratch, a daunting task even in the kindest of worlds. Instead, according to this view, the infant already has a store of basic concepts upon which to ground cognition. Learning does not consist in constructing concepts, but in relating real-world experience to the preexisting concepts innately provided. That is, the infant would learn by realizing that some preexisting concept—or combination of concepts built up from simple concepts by Boolean operations (AND, OR, NOT, and perhaps function applications, as with CAUSE and DIE for *kill*)—applies to some perceived object or event. Just as Plato's soul would learn by realizing there is a correspondence between an imperfect real-world object and a pure Form recalled from a trip to Platonic heaven, so the infant would learn by realizing there is a correspondence between a real-world percept, transduced by his senses to a mental representation, and a concept in Mentalese, recalled (as it were) by the ancestral memory in his genome.

Brains in SUVs

Methodological solipsism, along with the computer metaphor and the nativism that solipsism engenders, is very much a theory of late-stage capitalism. It is no surprise that methodological solipsism, as well as the whole analogy between computers and human cognition, arose and held sway starting in the last half of the twentieth century. Actual mechanical computers came into widespread use at that time, and a consumer-oriented market arose that placed primary emphasis on the self—the idea that prosperity and security would emerge from the satisfaction of individual wants. The market, by allowing individuals to think solipsistically, solely in terms of themselves, would transform individual self-interest into a greater common good; isn't that, after all, what Adam Smith argued in *The Wealth of Nations*? The result has been an emphasis on the self at the expense of greater social participation.

It's as though we are simply brains in SUVs driving, ever alone, in pursuit of our individual goals. We seek always to satisfy our individual wants and needs without knowing anything at all about the other SUV drivers. When we get to the big-box stores out in the suburbs, we busy ourselves like so many ants, carrying our purchases about and, in the process, generating the ant hill of our economy. Of course, the individual drivers don't have the slightest clue about their individual or cumulative effect on the world.

Certainly, computers and computation theory are compelling and profoundly useful in thinking about cognition and behavior, and I don't intend to throw the whole framework out. We can't do without symbols and operations on symbols. But I will argue that solipsism is an inadequate model for cognition and that an overemphasis on individual psychology has actually hampered our understanding of language. In this chapter and the next I argue that I am capable of meaning things—my words and mental states have content in the real world—precisely because I am part of a broader social network that gives my words and mental states content. The sentences on this page have content because you and I together are engaged in constructing meaning. By myself, I can't mean anything. Instead of rejecting the entire computational metaphor, I propose that we escalate from a computer metaphor to a network metaphor; where methodological solipsism thinks of a Univac mainframe, I propose that we think in terms of the Internet.

Symbols and Proofs

Let's first consider some limitations to solipsism. These are not reasons, by themselves, to reject solipsism and the computational view, but they should give the solipsist pause.

The first observation is that the computational/solipsist viewpoint is *formal*, as Fodor notes. This means that the symbols in the computational system are manipulated by the operations of the system without regard to what those symbols denote in the world. Once something has been encoded as formal symbols, the system will simply operate on the symbols according to rules that are blind to what the symbols are supposed to mean. Of course, we have to be very careful about how we use the symbols to correspond to meaning, lest things go fatally wrong. This approach relies on a branch of mathematical logic called *proof theory*.

Proof theory is an ingenious method of transforming semantic problems into problems that can be solved purely by symbol manipulation. To illustrate the idea, let's return to the well-known Aristotelian syllogism:

All men are mortal.
Socrates is a man.
_____
Therefore, Socrates is mortal.

Remember that the sentences above the line *entail* the sentence below the line. Proof theory offers a way to check this entailment without having to worry about the semantic content of the sentences. What we want to do is translate the sentences into symbol sequences. First, let's use the symbol ∀ to mean 'every' and the symbol → to mean something like 'if such and such is true, then something else is true'. We might then formalize the first premise, *every man is mortal*, as

$\forall x[\text{MAN}(x) \to \text{MORTAL}(x)]$,

which would mean something like 'for everything in the universe, call it x, if x is a man then x is mortal'. The x is called a *variable*, and it is basically a placeholder to keep track of where things occur in the sentence. A little reflection should confirm that this formula means much the same thing as *all men are mortal*. Similarly, *Socrates is a man* could be represented by the string

MAN(Socrates).

The trick in proof theory is to give rules that will allow anyone to manipulate the symbols (even if they don't understand what the symbols mean) in such a way that they come to a valid conclusion. For example, one might have the following proof rule:[1]

(10) **Proof Rule I**
   Given an expression $\forall x S[x]$, choose any name, erase the $\forall x$ part, and replace the variable x by the chosen name.

In (10) the expression $S[x]$ means a formal expression containing one or more occurrences of a variable x. So, given the expression

$\forall x[\text{MAN}(x) \to \text{MORTAL}(x)]$,

I can choose any name I like, for example, Socrates, and write

[MAN(Socrates) → MORTAL(Socrates)].

The next proof rule might be something like the following:

(11) ***Proof Rule II***
Given an expression $P \to Q$ (where P and Q are expressions) and given also the expression P, write the expression Q.

Proof Rule II says that if given

MAN(Socrates) $\to$ MORTAL(Socrates)

and also

MAN(Socrates),

then we may write

MORTAL(Socrates).

The result is that analogous to the syllogism, we get the following proof:

1. $\forall x[\text{MAN}(x) \to \text{MORTAL}(x)]$ Premise
2. MAN(Socrates)   Premise
3. MAN(Socrates) $\to$ MORTAL(Socrates)   1; Proof Rule I
4. MORTAL(Socrates)   2, 3; Proof Rule II

Each line of the proof is numbered. A line may be introduced either as a premise, which is noted on the right, or as a result of a proof rule. If a new line is entered because of a proof rule, then the name of the proof rule is given on the right, together with the numbers of the lines in the proof that the proof rule used.

The idea is to take any argument and translate it to a formal expression. The proof rules are simply manipulations on the formal expressions, without regard to their content. A person (or machine) armed with the proof theory could then go about constructing valid arguments without knowing what any of the arguments mean.

Proof theory is probably one of the greatest contributions of modern mathematical logic. The early pioneers of proof theory—Frege, Gödel, Church, Turing, and others—laid the foundations of computer science. The contribution of their ideas to cognitive science are evident in the quotes from Fodor. As it happens, we have a *complete* and *consistent* proof theory for some simple but powerful logics. What do I mean by *complete* and *consistent*?

A proof theory is complete if every valid semantic argument (like the preceding Aristotelian syllogism) has a corresponding proof. That is, if some collection of statements entails a particular sentence, then a complete proof theory will provide a corresponding proof by translating the premises into formal expressions, manipulating them with the proof rules, and transforming them into an expression that can be translated back to the conclusion.

A proof theory is consistent if the manipulations of the expressions dictated by the proof theory never output an expression that would be translated back to a false conclusion when the premises of the argument are true. If a proof theory is not consistent, it is basically useless. We couldn't trust any of the proofs it built because the conclusion might be false.

Now, it happens that an important branch of logic—first-order logic, which corresponds to reasoning with "and," "or," "not," "if...then...," "every," and "some"—has a complete and consistent proof theory. Any argument in first-order logic can be formalized, and if the argument is correct, we can eventually produce a proof of that argument, even if we don't understand what the symbols mean; we just have to obey the rules of proof. It might take an extremely long time to produce a proof, but we're guaranteed that if a proof exists, we will eventually find it.

This is all very impressive and important from the point of view of computer science. The fly in the ointment is that many logics don't have a complete and consistent proof theory, and as a matter of mathematical certainty, they never will. First-order logic, for example, cannot express the notion of infinity, nor can it help us reason about concepts like "most," as in

(12) Most dogs bark.

Many interesting areas of human reasoning do not have complete and consistent proof theories that simulate them.

Some of my colleagues are untroubled by this. Consider the following sentence:

(13) Most integers are not multiples of five.

At first glance, it may look like this sentence is plausibly true. In fact, though, it's false; there are just as many integers that are multiples of five as aren't. This can be demonstrated by setting up a function that, for every integer, produces a multiple of five: given an integer, multiply it by five and the result is a multiple of five. This shows that there are just as many multiples of five as there are integers.

Some would say that this shows that human reasoning about "most" is defective; presumably, it sometimes breaks down when we reason about infinite sets (such as the set of all integers). Thus, the proof theory that simulates reasoning about "most" might be incomplete. They argue that cognitive scientists should be working on fragments of proof theory for the more exotic logics. These fragments would be consistent but incomplete, and many truths might lie outside the proof theory. The incompleteness, of course, would have to model human errors in reasoning.

Now, it seems to me that this kind of argument makes a promise that is not yet fulfilled: a promise to produce proof fragments for these more exotic areas of human reasoning. But no one has yet produced a plausible candidate. I'm not suggesting that researchers stop trying to do this, but I think we should start looking at other techniques.

Perhaps some day someone will produce a needed fragment of the proof theory. It would simulate human reasoning in such a way that it works when we do and breaks down when we do. If someone did that, they could then program a computer or produce some other mechanical device that would appear to reason like a human. Would it follow that the machine was, in fact, reasoning like a human?

In an influential paper, "Computing Machinery and Intelligence" (1950), Alan Turing argued that a successful simulation of intelligence was in fact intelligence. At the core of the paper was the Turing test. Suppose someone claimed to have successfully simulated human intelligence with a computer. How would an impartial but critical observer judge whether the simulation was successful?

The idea would be to put a computer running the simulation in one room and a human being in another room. Messages could be passed back and forth from the rooms to the judge. The human being is instructed to be completely truthful, while the simulation can lie as much as its creators please. The judge is critical and allowed to ask any question she likes. After some number of questions, the judge must decide which room contains the computer and which room contains the human being.

For the test to be meaningful, the test must be run over and over again. Suppose the judge can only correctly guess half the time which room contains the human being. This would show that the judge was really guessing at random and could not distinguish the person (whose intelligence is not in question) from the machine (which is simulating intelligence).

The leap comes in supposing that a successful simulation of intelligence is in fact intelligence.[2] No one supposes that a good simulation of the weather is actually weather. In fact, it's quite reasonable to question whether a simulation that passed the Turing test would be the same in kind as genuine human intelligence.

Into the Chinese Room

The philosopher John Searle proposed a scenario that gives an alternative way of thinking about the Turing test. Searle (1984) invites us to imagine a man locked in a room. All he has in the room with him is a large book and a supply of pencils and paper. Periodically, a slip of paper is slipped under the door of the room. The paper is covered with what Searle calls "squiggles and squoggles," to which the man assigns no particular meaning. The man takes the paper and, following the instructions in his book, makes a series of "squiggles and squoggles" on his own paper. These new markings are based on the markings on the piece of paper slipped under the door and the instructions in the book and nothing else. The man can use the pencil and paper to work out the new sequence, perhaps erasing his markings and perhaps using scratch paper to finally work out a new sequence. Once he has finished following the instructions, he slips the resulting piece of paper under the door.

Unknown to the man in the room, the "squiggles and squoggles" on the pieces of paper slipped to him under the door are actually questions in Chinese, a language the man does not know. The pieces of paper that he produces are answers, in Chinese, to the questions. In fact, the man in the Chinese room is part of a Turing test. The whole Chinese room is analogous to a computer running a program. The slips of paper slipped under the door correspond to the input and output of the computer. The man is the central processing unit, the book is the software, and the paper and pencils are the memory of the system.

Suppose the Chinese room passes the Turing test, that is, a critical judge cannot accurately distinguish the output of the Chinese room from the responses of a native speaker of Chinese in some other room. Does it follow, asks Searle, that the man in the Chinese room speaks Chinese? The intuition is quite clear that although the whole system of the Chinese room appears to speak Chinese, the man in the room certainly does not.

Some researchers complain that Searle actually asked the wrong question. Clearly, the man in the room doesn't speak Chinese, but he's only one component in a larger system. Does the entire Chinese room speak Chinese, then? Some artificial intelligence and cognitive science researchers have argued that it does. If the room were equipped with a sufficiently large database (for example) or some other computational apparatus, then it would genuinely come to speak Chinese. The problem, though, is that nothing internal to the Chinese room or the symbols in the database connects the symbols to their referents. Adding more symbols and rules won't solve this problem, since the rules and symbols do nothing to connect the "squiggles and squoggles" to what they denote. All it will do is make the illusion of meaning more compelling for an external observer. In particular, a syntactic object like a database stored in a computer's memory will do nothing to endow the Chinese room with real semantic content.

Others have argued that Searle somehow got the whole scenario wrong and that the Chinese room is different in kind from a real computer. I'm not sure I see what the fundamental difference could be. But I think it's safe to say that whatever your ultimate position on the question is, there is some genuine discomfort at flat out saying that the Chinese room speaks Chinese. It just doesn't seem right to say that room really speaks Chinese, even if it always gives sensible answers to questions in Chinese.

Searle's analysis is that the Chinese room doesn't really speak Chinese because it lacks semantic content. Let's return to my trip to Prague in chapter 1. When I said that I wanted a plane ticket to Prague, I meant that I wanted a plane ticket—something that would allow me to get on a plane, not a train or a bus or a merry-go-round—and I meant Prague, not Poughkeepsie or Pittsburgh. My words had real content and by using those words, I intended them to have that content and no other content. Furthermore, by using those words, I intended that other people would grasp the content I intended. I wanted other people to understand what I meant; that was the whole point of saying particular things, after all.

As far as the Chinese room or the man inside the room is concerned, the squiggles and squoggles that are read are just squiggles and squoggles. The markings might mean 'tree' or 'Prague' or 'tiger' in Chinese, but neither the man nor the entire Chinese room has the slightest clue about what any of the markings mean. If this is correct, then methodological solipsism seems unhelpful if we want to understand how language has meaning or conveys information. It's fine to suppose that we are formal symbol-manipulating systems traveling around in skulls balanced on our shoulders, but ultimately human beings can mean things and perform feats of practical reasoning about the world based on those meanings, and that's what I want to understand. How is it that when I say "Prague" I mean Prague?

This capacity to mean something in particular by using a symbol is *intentionality*. Searle argues that people are different from computers in that we are capable of intentionality, whereas computers are not. Presumably, this is because of some biological fact about our brains. In other words, Searle claims that brains are capable of intentionality, whereas circuits and silicon are not. But here we return to my point that we often study what we find most puzzling. Why should brains have intentionality but other things not? I'm puzzled because I'm not sure I understand how a thing like a brain can have intentionality, whatever that is. We need to determine exactly what intentionality is.

Let's first ask whether brains are automatically endowed with intentionality. Is intentionality like some kind of neurotransmitter that brains automatically have but machines don't? Suppose some mad scientist had scooped out my brain at birth and kept it alive in a vat, feeding it a virtual reality simulation of the world. I seem to experience a tree and a bird and a snake, but really I'm experiencing a mental representation of a tree and a mental representation of a bird and a mental representation of a snake. Can I mean snake when I say (or experience the mental representation of saying) "snake"? Suppose the mad scientist had tricked me by getting me to associate a (mental representation of a) fire hydrant with the word *tree* and a (mental representation of a) penguin with the word *snake*. A sentence like

(14) The snake is in the tree.

might wind up meaning something like

(15) The penguin is on the fire hydrant.

if it meant anything at all. So it seems to me that intentionality doesn't come automatically with being, or having, a brain. I suspect that the roots of intentionality lie in our ability to coordinate our behavior socially. Chapter 4 looks at coordination games in more detail, and chapter 9 discusses the relation between coordination games and concepts.

I was once mulling over the Chinese room puzzle with a colleague in neuroscience. He had the very firm intuition that computers could have intentionality. He argued that his computer had a function that told him whether the printer he was using was low on paper or toner; if the printer was running out of toner, for example, he got a message telling him so. He thought that the computer had the property that it could denote the printer and the supply of paper or toner and that this proved computers were capable of intentionality.

The Social Nature of Intention

About a year after my colleague in neuroscience made the printer argument, it occurred to me where the trick lay. When I'm trying to print and I get an error message telling me that my printer needs more paper, it's not the machine that intends to tell me that; it's the programmers and the hardware designers who built the machine. The machine just "knows" that there's current coming from some sensor or other. It doesn't know a thing about paper supplies or toner. The designers of the machine—the hardware and software—constructed the machine to give a particular signal under certain physical conditions.

I'm able to interpret the machine's message because I'm keyed to the intentions of the designers. When the message comes up that the printer is low on paper, I think, "Aha! the machine is low on paper," and I behave accordingly. The machine seems to have intentionality because it is in a social network with human beings: me and its designers. If you take away the intentions of that social network—if you eliminate me and the programmers and the hardware designers—then the machine wouldn't "mean" anything even when it is physically in the condition of receiving current from whatever sensor indicates low paper. Intentionality is a social property that is enforced and given content by a social network of human beings.

Algorithms, formal manipulations of symbols, don't mean anything until they are endowed with intentions by software designers and users. I can use the same algorithm to sort a list of numbers or to alphabetize a list of words. The algorithm doesn't care as long as it's carrying out formal operations on symbols that those operations are defined on. As Fodor points out, the content of those symbols does not matter to the algorithm.

If a social network can endow a machine with intentionality, maybe the same thing is going on with my brain. That is, I'm capable of meaning Prague by *Prague* because I'm part of a vast social network of beings who use *Prague* to mean Prague. Intentionality is still a biological fact in the sense that we have evolved as social beings. As such, we enable each other to mean things. Any account of how we use language to mean things will have to grapple with the social side of meaning. Try as we might, we can't simply reduce meaning to the internal states of a machine or a brain; we need each other if we are to mean things.

The Excesses of Youth

As an undergraduate, I became obsessed with the idea that there could be a *philosophical anthropology*, a phrase I may have picked up from

Wittgenstein's *The Blue and Brown Books*, where he presents the idea of a *language game* as a tool of analysis. Looking back, I think I completely missed his point, but I was convinced that there was a method of combining ethnography of language with logical analysis. It seemed to me that linguistics was the obvious locus for this kind of research. Accordingly, I divided my time between Carnap's book *The Logical Structure of the World* (affectionately known as the *Aufbau*), Wall's textbook on mathematical linguistics, and Brown and Levinson's *Politeness: Some Universals in Language Usage*.

The Brown and Levinson book was exciting because it seemed close to the notion of a language game, a strategic game where players seek to maximize utility (expressed in terms of face). The work was not formal in the sense that I wanted it to be, but it was empirically precise. I thought I could see in it the outlines of the kind of formal ethnography of language that I was after. Wall's textbook was an introduction to the kind of formal systems that I thought would be useful. I found it a very comforting book, and my original copy was soon red with underlining.

Carnap's *Aufbau* was the model of what I wanted. In it, he tries to create an axiomatic system—a proof theory—that would reduce physics (and, I thought, the whole of physical experience) to a few axioms and some primitive relations. I found this a tremendously attractive project. I still love the idea of a fully constructive system. The idea would be to construct results—results about anything, say, about psychophysics or numbers—from a finite set of primitive propositions, along the lines of Euclid's *Elements*. At the time, I was also very much taken with the work of Nelson Goodman. His *The Structure of Appearance* was an attempt to repair the project in the *Aufbau*. I found his arguments in *Fact, Fiction, and Forecast* utterly intriguing and absolutely opposed to the idea of Mentalese. I think I'm only now coming to understand Goodman's constructivism.

When I went to graduate school, the big excitement was over Noam Chomsky's *Lectures on Government and Binding*, which had the look of the future about it. It seemed to me that the project was to derive grammatical systems from a small set of axioms. The exact nature of the axioms could be modulated by a set of parameters that would vary from language to language, but the underlying system would be fundamentally the same across languages. By specifying the parameters in a certain way, you would construct (or so I thought) the grammar of English; specify the parameters in another way, and you would get a totally different grammar for another language.

At birth, the learner would have the fundamental axioms along with the parameters—the potential for any possible language—prespecified; this innate equipment would be modulated by experience to yield the adult grammar. Being a constructivist, I misinterpreted a lot of the project, which was not really constructive. Nevertheless, I thought it could be made to be constructive, a kind of Euclid for the world's languages. Crucially, I thought, the essence of the system was internal to the individual.

The combination of this line of thinking with the Tarskian method of semantic analysis was exactly what I needed. Fodor's methodological solipsism fit naturally into this project. On this view, meaning would arise from internal properties of mental representations, although it would be modulated by the individual's experience in the world. It all seemed to have an inexorable logic to it. And, of course, it put the focus on the self, something which pleased my youthful egotism.

I shied away from the question of where meaning came from, how these mental representations acquired meaning. I read Searle but tried to dismiss his argument as a misunderstanding, as though Searle were talking at cross-purposes to what I was doing. But I started to feel uneasy.

My uneasiness only worsened when I read Putnam's "The Meaning of 'Meaning'." His argument that meaning was not a property of the individual but, rather, a social property couldn't be avoided. The idea got into the back of my brain and began to dig in. I couldn't dismiss the arguments easily, and the more I thought about them, the more uneasy I became. Eventually, my fall from Platonic heaven was complete.

In chapter 3, I flesh out the social side of meaning a bit more. Then part II presents game theory as a way of investigating linguistic meaning. The first approximation of meaning involves purely competitive games with a clear winner and a clear loser. Part III suggests that meaning involves real cooperation and examined game-theoretic models of cooperative meaning.

Further Reading

Phrase structure grammars (and many other things) are discussed in an approachable way by Partee, ter Meulen, and Wall (1990), although I have a sentimental attachment to Wall (1972), which I read when I was twenty and lived in Austin, Texas.

The discussion of grammar and compositionality is an amalgam of various sources. See, in particular, the discussion of Montague grammar in Partee, ter Meulen, and Wall (1990). The "hat" notation is a simplified version of λ-abstraction, for example. Of course, I'm combining this with a variety of notions from the literature on Mentalese, but I don't think I'm getting anything wrong here.

The Allegory of the Cave is from Plato's *Republic*. I was teaching an introductory linguistics class once, in California, and the subject turned to innate ideas. I described the thinking, and a student approached me after class and said that she didn't believe in innate ideas but she didn't think that people learned language from experience either. When I asked her how people acquired language, she replied that they simply remembered languages from past lives. This is actually pretty close to Plato. When I asked her what she did for a living, she told me she was an assistant fire walker, and she had to go prepare the coals. An interesting conversation.

The idea that representation somehow explains meaning is deeply embedded in both linguistics and cognitive science. Rorty (1979) is a signal work questioning these theories. Certainly, linguists and cognitive scientists would do well to consider the alternatives he advances. Although I am not in total agreement with him, I find his arguments interesting and compelling.

A good source for proof theory is Barwise and Etchemendy (2002). I'm also fond of Bostock's (1997) book, which does a lovely job of introducing a variety of proof theories. Landman (1991) gives a nice proof that the determiner *most* cannot be expressed in first-order logic.

The Turing test is discussed in Turing (1950), which has been widely anthologized and is justly famous. The Chinese room problem can be found in Searle (1984), and a large literature has grown up around it. Noë (2009) argues that consciousness itself arises from the interaction of the brain with the external world. As it is for consciousness, so it is for meaning.

The notion of language game can be found in Wittgenstein (1953; 1958). Particularly important to my thinking about this is Kripke's (1982) commentary on the private language argument. Kripke is right to emphasize the social nature of rule following, and with Putnam's (1975) paper, Kripke's work served to undermine my confidence in solipsism.

Putnam's (1975) paper is worth seeking out. Putnam (1981) has also been quite influential and had a big impact on my thinking about language. The papers collected in Pessin and Goldberg (1996) are very useful. I return to some of Putnam's ideas in chapter 9, in particular, giving a social account of word meanings using Schelling's (1960) notion of focal point and relating it to prototypes (Rosch 1978; Murphy 2002).

3   Meaning and the Social Contract

Choice and Meaning

In Lewis Carroll's *Through the Looking Glass,* Alice meets Humpty Dumpty, a rather irascible fellow, and they fall into a discussion of the difference between birthdays and unbirthdays. Humpty Dumpty argues that unbirthdays are clearly superior, since there are 364 days for unbirthday presents:

"And only *one* for birthday presents, you know. There's glory for you!"
   "I don't know what you mean by 'glory'," Alice said.
   Humpty Dumpty smiled contemptuously. "Of course you don't—till I tell you. I meant 'there's a nice knockdown argument for you'."
   "But 'glory' doesn't mean 'a nice knockdown argument'," Alice objected.
   "When *I* use a word," Humpty Dumpty said in rather a scornful tone, "it means just what I choose it to mean—neither more nor less."
   "The question is," said Alice, "whether you *can* make words mean so many different things."
   "The question is," said Humpty Dumpty, "which is to be master—that's all."

Humpty Dumpty is right that *choice* is fundamental to how we are able to mean things using words. The great Swiss linguist Ferdinand de Saussure wrote,

Our memory holds in reserve all the more or less complex types of syntagms [linguistic units that are in a sequential relationship to one another], regardless of their class or length, and we bring in the associative groups to fix our *choice* [emphasis added] when the time for using them arrives. When a Frenchman says *marchons!* '(let's) walk!' he thinks unconsciously of diverse groups of associations that converge on the syntagm *marchons!* The syntagm figures in the series *marche!* '(thou) walk!' *marchez!* '(you) walk!' and the opposition between *marchons!* and the other forms determines his choice; in addition, *marchons!* calls up the series *montons!* '(let's) go up!' *mangeons!* '(let's) eat!' etc. and is selected from the series by the same process. In each series the speaker knows what he must vary in order to produce the differentiation that fits the desired unit. If he changes the idea to be expressed, he will need other oppositions to bring out another value; for instance, he may say *marchez!* or perhaps *montons!* (Saussure 1960, 130)

Saussure was bringing out the idea that units have meaning to the degree that there is a choice between them; I can't vary meaning unless I'm able to choose different units. That, I think, is his notion of opposition, and it is explicated quite nicely by game theory.

As usual, though, Alice is the sensible one. We can't make words mean what we choose them to mean. Try as we might to master words, words always master us, because words are backed up by years of social habit, conventions built up over time and supported by a whole community of speakers whose tacit agreements about the conventional meanings of words gives those words their particular contents. Without the backing of that community of speakers, we couldn't mean anything with words. Humpty Dumpty can't make a word mean whatever he chooses it to mean any more than I can make drivers go at red lights and stop at green lights; social conventions are against us.

Internal Predicates and External Behavior

Even the most adamant proponent of Mentalese will admit that some aspects of language must be learned from the external world and that language has a social component. Languages, after all, vary one from the other in how they encode meaning. A child learning a first language must be, in part, socially conditioned; children must observe how the adults around them use the language and discover the correspondence between what is said and what is meant.

Let's imagine that we've managed to cull out all the predicates of Mentalese. That is, we've found the concepts around which the Language of Thought organizes itself. This means that, for any natural language, we have the basic ingredients that will allow us to express the meaning of any word in that language as some combination of predicates from Mentalese (see chapter 2). For example, *kill* is made up of a kind of compound of CAUSE and DIE. It happens that languages vary as to how they encode concepts. To take a famous example, Spanish encodes how someone undergoes (but does not cause) an event differently than English does:

(1) Se me   quebró el   brazo.
    self to me broke   the arm
    'The arm broke itself to me.'

The first line in (1) is the sentence in Spanish, the second line is a translation of each word into English, and the third line is a direct translation, putting aside questions of usage. The most colloquial way to say the Spanish sentence in (1) would, of course, be

(2) I broke my arm.

Notice how Spanish and English express the meaning using different syntactic forms. Spanish has *el brazo* ('the arm') as the subject of the sentence, whereas English has it as the direct object of the verb meaning BREAK. English has the person undergoing the breaking as subject, whereas Spanish encodes it as an indirect object of the verb. Finally, Spanish uses a reflexive (*se* 'itself'), which English doesn't use at all.

In other words, if Mentalese exists, then languages can differ as to how they express its predicates. Any linguist can list numerous cases of how one language will express a meaning differently than some other language. Even within a language, we have a variety of choices for how a meaning is expressed:

(3) a. Big Tony twisted my arm.
    b. What Big Tony did was twist my arm.
    c. My arm got twisted by Big Tony.
    d. He twisted my arm, Big Tony.

All the sentences in (3) express the same proposition, namely, that Big Tony twisted my arm, but the different forms carry different meanings. Suppose you see me with my arm in a sling and ask,

(4) What happened to your arm?

I answer,

(5) My arm got twisted by Big Tony.

In this case, the flow of conversation is connected from sentence to sentence. The topic of the conversation is my arm and what happened to it, and my answer makes *my arm* the subject of the sentence. But if I answer,

(6) What Big Tony did was twist my arm.

you would think me quite peculiar. We were talking about my arm, not about what Big Tony did. On the other hand, if you see Big Tony in handcuffs and ask what he did to get arrested, I could answer with (6), and in that case (5) would seem peculiar.

We have all learned facts about how meaning is encoded in a particular language we speak—knowledge we share with other speakers of the language—that we could not have been born knowing. Instead, we acquired this knowledge through intense social practice. We know how to signal particular meanings and how to meet particular social ends. Even the most ardent Mentalese theorist will agree that there is a social component to language. She might try to simulate a child's social environment by imagining a machine being fed a text of sentences which might be "annotated" to show communicative intent and facts about the social context, but this, too, would be a tacit admission of the social face of language.

Public Knowledge

Once we acknowledge this social side to language, we have stepped on a slippery slope. Having set foot in the social, we tumble from the solipsistic serenity of Platonic heaven to the complex bustle of the social world. Now, I don't argue that there is no innate component to our ability to acquire and use language; surely there is. But language is limned with the light of others. Having seen that light, do we really want to retreat to the darkness of the cave?

Of course, the truly committed solipsist could argue that the entire social face of language can be simulated by adding logical operators of one sort or another to the vocabulary of Mentalese. I argue, though, that such a simulation would forever limit understanding of linguistic meaning. To see this, let's consider a puzzle renowned among game theorists. The puzzle exists in many versions, but I give my version as the *dirty frat boy problem*:

The dean of undergraduates at a certain Ivy League institution dropped by the Pi Upsilon fraternity house to borrow the boys' beer bong for an upcoming undergraduate council meeting. As part of an elaborate fraternity prank, all the reflecting surfaces in the frat house had been painted over with a matte black paint. As fate would have it, the boys all had dates that night with various sorority sisters over at the Sigma Lambda Tau house.

The boys, of course, wanted to impress the girls with their suave and debonair good looks. Alas, some of the boys had gotten their faces dirty while playing a particularly brutal game of Wiffle ball that afternoon. This being the Ivy League, the boys were very competitive; hence, the brutality of the game and the following rather sad comment on human nature: each boy secretly hoped that his own face was clean and that his fellow's face was dirty, for if his fellow's face was dirty, the original boy (if clean) could go on his date looking good by contrast.

So no boy could get a reliable answer from any other boy if he asked whether his own face was clean. And, of course, no boy could look in a mirror to verify that his own face was clean. Furthermore, there is some risk that wiping one's own face wouldn't clean it. What to do?

Into this sad situation comes the dean of undergraduates on his quest for the beer bong. He is told of the prank with the matte black paint.

The dean takes pity on the boys' plight. Being an academic, his weapon of choice is the Socratic method. To the eager assemblage of Pi Upsilon frat boys, he says, "At least one of you has a dirty face. Do any of you know if you have a dirty face?"

The boys reply in unison, "No!"

The canny dean then asks, "Now, do any of you know if you have a dirty face?"

Once again, the puzzled boys reply as one, "No!"

The wily dean repeats his question: "Now, do any of you know if you have a dirty face?"

Enlightened, some of the boys say, "Yes!"

"Glad to be of service, lads," says the dean, raising the beer bong in salute and stepping out into the night.

Now comes the puzzle: How many boys said "Yes!"?

In order to solve the problem, first consider the case where there is only one dirty frat boy. He looks at his fellows and notes that they have clean faces. When the dean announced that at least one boy had a dirty face, this would have been informative to the lone dirty frat boy; seeing the mass of clean faces and no dirty face among them, he would think, "It must be me who is dirty. Otherwise, the dean would have said something false. I have learned that I have a dirty face." Then, when the dean asked again if any boy knew whether he had a dirty face, the lone dirty frat boy would say "Yes!" and that would settle the matter.

Now suppose there are two, and only two, dirty frat boys. One of them looks around and sees one, and only one, dirty frat boy, his colleague in grime. Of course, neither dirty frat boy (nor any of their fellows) knows that he himself is dirty. Now the dean announces that there is at least one dirty frat boy. All the frat boys look around. The dirty frat boys see one other dirty frat boy. The clean frat boys see two dirty frat boys. They all think to themselves, "The dean has told me what I already know. I can see at least one dirty frat boy. I am in the dark about myself though." Now when the dean asks his question, all the boys answer "No!"

But now consider the perspective of a dirty frat boy. He sees only one dirty frat boy and says to himself, "Wait, I can see only one dirty frat boy. If there were only one dirty frat boy, the one I see would surely have answered "Yes!" since the dean's statement would be news to him.

I conclude that there must be another dirty frat boy somewhere. I see by inspection only one dirty frat boy. Therefore, I must be the second dirty frat boy." When the dean asks his question again, both dirty frat boys will reply "Yes! I know I'm dirty."

If there is only one dirty frat boy, the dean needs to ask his question only once and the dirty boy will know he is dirty; if there are two dirty frat boys, the dean needs to ask his question twice; after the second round, the dirty boys know they are dirty.

Suppose there is some number, n, of dirty frat boys. If there is one dirty boy, he will look around and see $n-1$ dirty boys, since he can't see himself. When the dean asks his question $n-1$ times, he will reason, "Wait! If there were only $n-1$ dirty boys, then the dirty boys should have answered 'Yes!'. They didn't do so. The only explanation for this is that there are $(n-1)+1 = n$ dirty boys here." Thus, when the dean asks the question for the nth time, the one dirty boy will be among the boys who reply "Yes!"

Notice that a clean frat boy won't know he's clean until the dirty frat boys answer "Yes!" The reason is that he observes there are k dirty boys, but as far as he knows, it's entirely possible that there are $k+1$ dirty boys. One of the dirty boys, however, will see only $k-1$ other dirty boys; at the kth time, the dirty boys will know they are dirty, but a clean frat boy will still be in doubt.

Why is this interesting? The example shows that there is a difference between public knowledge and private knowledge. In the case where there are more than one dirty frat boys, the dean never says anything that the boys don't already know: each can see that there is at least one dirty boy. At each round of questions, though, each boy makes public that he doesn't know the answer. When this information becomes public, the inferences the group can make from that information change.

The really interesting thing is that all the boys are in doubt about whether they are dirty. Everyone in the group knows this, but when it becomes common knowledge—everyone knows that everyone is in doubt, and everyone knows that everyone knows this—the information becomes useful in a particular way that it wasn't useful before.

I can now state an obvious property of language:

(7) Natural languages allow private information to become common knowledge.

By common knowledge I mean that everyone in some group knows the same thing, they know that all of them know it, they know that they know that all of them know it, and so on. Game theory has the interesting property of explicitly representing common knowledge in the form of a game. Game theory allows us to step outside our heads and reason about social knowledge as social knowledge.

Now, someone could argue that the solipsistic theory can easily solve the dirty frat boy problem. The only requirement is to add logical operators. One operator might be something like $K_i\phi$, denoting 'agent i knows that ϕ is true'. Another operator might be $E_{\{i,j,k\}}\phi$, denoting something like 'agents i, j, and k each know that ϕ is true'. That is, they individually know that ϕ is true, although they might not know that the other agents also know that ϕ is true. Finally, an operator $C_{\{i,j,k\}}\phi$ would be taken to mean that it is common knowledge among the agents i, j, and k that ϕ is true.

A solipsist would let ϕ be "I am dirty" and then try to write a logic where no individual frat boy, x, knows he's dirty:

$\forall x \neg K_x \text{DIRTY}(x)$.

If there are more than one dirty frat boys, then the dean's announcement,

$\exists x \text{DIRTY}(x)$,

is not informative. Each frat boy has a private information state—a set of propositions that he knows for a fact—that contains the dean's proposition.

As each frat boy, i, announces that he doesn't know whether he's dirty,

$\neg K_i \text{DIRTY}(i)$,

no one learns anything new. The solipsist would then need to derive the fact that each frat boy knows:

If there are $k - 1$ dirty frat boys, they will learn that they are dirty at the $(k - 1)$th answer to the dean's question.

Furthermore, they each should know the following:

If the $k - 1$ dirty frat boys I see answer yes to the dean's question, then I'm clean.

And so on.

The solipsist, in other words, has to reproduce the external social situation inside each frat boy's mind. For the solipsist, the only thing that can give force to propositions that each frat boy considers is the role the symbolic representation of that proposition plays in a larger formal system. For each frat boy, the dean and all the other frat boys need not exist.

All that matters are the formulas and whatever system of formal manipulation is set up to operate on the symbols. Once again, here is an attempt to reconstruct Platonic heaven inside the cave.

The Economics of Meaning

Is the solipsistic program adequate to the task of social cognition? I suspect not. To consider why, I look at a bit of the real world, an example concerning the value of money in a real economy. The economy is in Yap, a small group of islands in the South Pacific, and one of its unusual features is that its money is made of stone.

The Yapese have used large stone wheels called *rai* (figure 3.1) for several centuries. These stone wheels are made from a kind of limestone that is not indigenous to Yap but is quarried from the Palau islands, some 250 miles to the southwest of Yap.

The wheels themselves vary in size. Some are only a few inches in diameter, and others reach a diameter of 12 feet and weigh thousands of pounds. The *rai* have no intrinsic value as anything other than money. They are not ornamental to the Yapese, they are not functional—they aren't used as wheels to move things, for example—and they apparently have no spiritual significance. They are simply money. Since they have

Figure 3.1
Rai (Stone Wheel) from Yap

no intrinsic worth—they aren't themselves valuable as commodities, and they don't represent some other valuable thing like gold—the Yap stone money is a prime example of fiat money (along with the U.S. dollar), that is, it is simply declared to have value by a central authority.

The fact that the *rai* are such pure fiat money makes them rather like signs in Saussure's sense. Saussure thought of a sign as an arbitrary link between a signifier (for example, a word like dog) and a signified (the actual thing that *dog* can denote). There is no necessary connection between *dog* and dog. I could use *pupperino* to refer to a dog. But I don't because there's a certain amount of social entrenchment among English speakers on the side of using *dog*. It would take a lot of work to get everyone to start using *pupperino* instead. The arbitrary link between signifier and signified is socially enforced.

Saussure was prescient in seeing the relation between meaning and economics:

To determine what a five-franc piece is worth one must therefore know: (1) that it can be exchanged for a fixed quantity of a different thing, e.g., bread; and (2) that it can be compared with a similar value of the same system, e.g., a one-franc piece, or with coins of another system (a dollar, etc.). In the same way a word can be exchanged for something dissimilar, an idea; besides, it can be compared with something of the same nature, another word. Its value is therefore not fixed so long as one simply states that it can be "exchanged" for a given concept, i.e., that it has this or that signification: one must also compare it with similar values, with other words that stand in opposition to it. Its content is really fixed only by the concurrence of everything that exists outside it. Being part of a system, it is endowed not only with a signification but also and especially with a value, and this is something quite different. (Saussure 1966, 115)

So how does the Yapese stone money work? Is it really money? Normally, money works as a medium of exchange, a store of value, a unit of account, and a standard of deferred payment. The role of money as a medium of exchange is easy to understand; you buy things with money. Equally, money stores value in the sense that, if you save money, you have stored up value. Money as a unit of account involves measuring goods and services into monetary units; for this to work, money should be divisible into smaller bits and fungible in the sense that the money is seen as equivalent in value to other goods and services.

The *rai* have value. In the late nineteenth century, a British naturalist reported seeing 400 Yapese men producing stones on Palau that would be transported back to Yap. From this, it has been estimated that 10 percent of the adult male population was in the business of producing the money. It is clear from this that the money had some kind of value;

otherwise why would the Yapese have devoted so much energy to its production.

How does it get its value, though? Bryan (2004) observed that the Yap chiefs authorized expeditions to acquire new stones. They retained all the larger stones and two-fifths of the remaining smaller ones, a kind of tax on the Yapese. In effect, the chiefs acted as central bankers for the island.

The individual stones were assigned a value, although the method for doing so is unclear. The size of the stone did not fully determine the stone's worth. The values varied depending on the expense and difficulty of bringing the stones back to Yap. Stones that involved peril, even loss of life, were most highly valued (which suggests that there may be a possible spiritual dimension to the stones' value for the Yapese). Stones that were cut using shell tools and carried in canoes were worth more than stones that were quarried with iron tools and transported by Western ships.

In fact, an Irish American from Savannah, Georgia, named David O'Keefe, was shipwrecked on Yap in the nineteenth century. He developed the scheme of quarrying the stones with modern tools and transporting them on ships. Once on Yap, he would trade his stones for coconut meat, which he could then transport to the West. During O'Keefe's time, the stones measured from 4 feet to 12 feet in diameter. In addition, the number of stones grew dramatically; some 13,000 were counted in the 1920s, although the stones were esteemed rare in the 1840s.

You might think that O'Keefe's project would result in inflation, with the individual *rai* being worth less. In fact, the older stones retained their value, and O'Keefe's *rai* had less value.

While the *rai* have value, it's harder to argue that they are a unit of account. The problem is that the *rai* are not divisible. You can't take a stone wheel and cut it in half to provide change. Indeed, most *rai* are worth quite a bit. In the early twentieth century, a 25-inch *rai* was taken as being worth fifty baskets of food or a full-sized pig. A stone the size of a man was taken as being worth whole villages and plantations. Thus, *rai* were normally used for large transactions. Smaller transactions were handled with barter or pearl shells. Zelizer (1997) noted, for example, that strings of mussel shells served as women's money.

Do the *rai* work as a medium of exchange? Most of the *rai* are so large that it isn't practical to transport them. They remain fixed at their physical locations. But they do change possession. Of course, the physical location of most dollars is irrelevant, too; they can change hands electronically. *Rai* aren't very good as a medium of foreign exchange. In fact, the U.S. dollar has been the legal tender on Yap since 1986, although *rai* are still used for some domestic transactions. Their fixed location and lack of intrinsic value limit their appeal to outsiders. Nevertheless, within Yap they seem to have retained their purchasing power.

So how do the stones work in practice? Bryan's examination of the record suggests that the stones act as markers. For example, suppose an islander wishes to fish in someone else's waters. He might pay the stone in recognition of this service. Once the fish are caught, the fisherman gives an appropriate number of fish to the owner of the fishing waters and reclaims the stone. Thus, the stones act as a kind of memory marker for economic exchanges. Occasionally, one group will exchange a stone outright with another group as a recognition of aid given by the latter group to the former. If the former group later gives aid to the recipients of the *rai*, it will be returned.

From a strictly utilitarian view, it might seem puzzling that so much effort and so many resources would be expended to support a system of markers—IOUs, if you will. After all, all you need for a marker is a piece of paper and a pen, or some other small token. Crucially, the marker must be public information, part of the common knowledge of the group; otherwise it might not carry any weight. It might be, as Bryan speculates, that the Yap chiefs lacked sufficient credibility to decree an object's value and thus needed an object to which value had been assigned and which could not be easily replicated. Thus, the markers would take on value from the social environment and would work as sufficient security to act as memory markers for meaningful transactions. As Zelizer points out, money in general has social meaning for the people who use it that goes beyond the utilitarian considerations of economics.

Physical Computation and Social Computation

Speaking of utilitarian considerations, let's visualize the island of Yap—the stone *rai* and all the people on the island—as a kind of computer. In this case, we take the island to be a computer for working out the economy of stone wheels, their value, and their distribution among the various possible owners. It's hard to do this thought experiment with the U.S. economy because so much of our economic activity involves agents from outside; we couldn't take the physical country, the United States, as a similar computer because of this. But the *rai* are of no use for foreign exchange, so it's at least a coherent thought experiment.

When we think about computers at the level of software, we can think of the computer as operating on internal representations. These internal representations, at least broadly speaking, can be construed as arising from the physical state of the machine; the internal representation is a consequence of the physical state of the machine. We might suppose, also, that the brain is like a computer in having a physical state upon which mental representations depend. In the case of the machine, the physical state is given by a description of the states of the individual on-off switches that make up the computer's internal machinery. In the case of the brain, we might suppose that the physical state is given by the levels of activation of the various regions of the brain.

What about the island of Yap? Does the economy arise from the physical state of the island? If the *rai* were more like coins—small, portable things—we might know by following the positions (that is, the physical locations) of the coins on the island. But the physical positions of the *rai* need not change in order for them to change possession. So, although the *rai* act as markers of memory—in particular, a memory of debt—their physical properties seem only indirectly relevant to their value or their current place in the economy. Thus, we can't follow the state of the Yap economy by looking at the physical state of the island.

A true believer in methodological solipsism would have to argue that the state of the island economy is, at least in principle, a function of the mental states of some individual on the island, or even of all the islanders. If so, we could gain all the information about the economy by "reading" the brain states of some or all of the individuals on the island. Now, in reality, it might be the case that the information is distributed over several individuals, but in principle it should be possible for one person to know all the crucial information about both the values of the individual *rai*—there are about 13,000 of them—and who on the island possesses each one.

The chiefs, for example, might be taken as having absolute knowledge of the values of the individual *rai*. The chiefs clearly have some power in setting values, and the credibility of the chiefs is crucial in maintaining the whole system. But credibility is an inherently social notion. The chiefs have credibility to the degree that members of the society believe they are credible. Suppose a chief went visibly and demonstrably mad and had delusional beliefs about the *rai*. The chief's beliefs would no longer carry much weight because he would no longer be a credible agent.

Even if the chiefs are credible about the values of individual *rai*, the possession of the *rai* is a matter of social knowledge. Clearly, no one person can or should have definitive exclusive knowledge about the possession of the stone wheels. Possession of the *rai* is a matter of social exchange, known to many people, and not a matter of the beliefs of an individual.

Of course, this isn't surprising. Economies are big systems, and one wouldn't necessarily suppose that an economy is like meaning, or so the solipsist might argue. While I have the robust intuition that I *know* what things *mean*, I don't have any such intuition about the economy. The economy is something that is certainly external to my mental state; economic things happen to me, and my contribution seems small indeed.

But how good is this intuition? Something gives value to currency and something gives value to words. In the case of currencies, their value is set by social practice; each day, there are untold millions of transactions that contribute to the overall computation of value. No one person's opinion or beliefs matter much to the system, although individuals might influence the outcome.

The Sociolinguistics of Meaning

Even if we suppose that words denote concepts in Mentalese, it still has to be the case that something enforces these values, just as is the case with currency. Unless we share the denotations of the words we use, at least for the most part, how could we ever transmit information to each other? Is there some private fact about me that makes *dog* mean dog? No, it has to be that my meaning conforms to external public usage. Just as Humpty Dumpty can't choose to make *glory* mean 'a nice knockdown argument', I can't choose to make *dog* mean anything other than what social practice dictates it can mean.

Let me illustrate this with a personal anecdote. I decided once that it would be a good idea to live in Europe, so I left my job in the United States and moved to Geneva, in the French-speaking part of Switzerland. As it happened, my French was dismal; I had some college French, which only remotely resembled what they spoke in Switzerland. Happily, I was working as a research scientist, so I didn't need much French, but I thought that it was only sporting to try to learn the local language.

My plan for learning French was based on immersion. I listened to France Info, the French news radio station. I came home in the late afternoon and watched reruns of *Alf* and *MacGyver* dubbed into French, and after dinner I read French novels. I preferred nineteenth-century novels and detective stories. The detective stories were straightforward, but the nineteenth-century novels were nightmarish. Well, to be perfectly truthful, the whole thing was nightmarish; one minute it was Alf and the next minute it was Gerard de Nerval. Small children were more articulate than I was, and I spent what seemed ages listening to the radio and watching television without understanding one word. I sweated for hours with various dictionaries, but the vocabulary often turned out to involve items that were useful back in the nineteenth century, things involving carriages and horses and ancient household paraphernalia.

Most discouraging was the names of flowers and plants. Since I was raised in the desert, I had never really learned much about botanical terminology aside from *greasewood*, *tumbleweed*, and a variety of cacti. So if, in my reading, I came across the word *prêle*, I would look it up and find it means equisetum, whatever that is. The word *gentiane* means gentian, but what is that actually? *Jonquille* refers to a daffodil, but all I knew about daffodils was that they were some kind of flower.

It suddenly occurred to me that I was simply exchanging one symbolic token for another without being able to connect it to an actual thing.

A few years after I returned to the States, I read Putnam's article "The Meaning of 'Meaning'" with some sympathy. In it, he confessed that he didn't know an elm from a beech. In fact, he pointed out, his concept of "elm" was the same as his concept of "beech, namely," some kind of tree. I'm not sure what the difference is either. But they certainly don't mean the same thing. We both know that elms are different from beeches. When I say "elm" I mean elm and when I say "beech" I mean beech. I know that they don't mean the same thing.

Putnam imagined a planet directly on the other side of the sun from us—we can never see it because the sun is always in the way—called Twin Earth. Everyone on Earth has a molecule-for-molecule identical twin on Twin Earth. The histories of Earth and Twin Earth are nearly identical.[1]

> If someone heroically attempts to maintain that the difference between the extension of "elm" and the extension of "beech" in *my* idiolect is explained by a difference in my psychological state, then we can always refute him by constructing a "Twin Earth" example—just let the words "elm" and "beech" be switched on Twin Earth.... Moreover, I suppose that I have a *Doppelgänger* on Twin Earth who is molecule for molecule "identical" with me (in the sense in which two neckties can be "identical"). If you are a dualist, then also suppose my *Doppelgänger* thinks the same verbalized thoughts I do, has the same sense data, the same dispositions, etc. It is absurd to think *his* psychological state is one bit different from mine: yet he "means" *beech* when he says "elm" and I "mean" *elm* when I say elm. Cut the pie any way you like, "meanings" just ain't in the *head*! (Putnam 1975, 144)

What makes Putnam mean elm when he says "elm"? It's that *elm* means elm in his language. Experts on trees agree: *elm* means elm. Similarly,

French experts on plant names agree that *jonquille* means jonquille, while English-speaking experts will say that this is just the French way of talking about daffodils. In meaning things, we all rely on the expertise of speakers outside of us to give our words their content.

Putnam called this the Hypothesis of the Universality of the Division of Linguistic Labor:

> Every linguistic community exemplifies the sort of division of linguistic labor just described: that is, possesses at least some terms whose associated "criteria" are known only to a subset of the speakers who acquire the terms, and whose use by the other speakers depends upon a structured cooperation between them and the speakers in the relevant subsets. (Putnam 1975, 145)

It's no wonder that Putnam's article resonated with me after my adventures with French flower terms; every word we use has behind it the weight of social practice. That social practice gives my words content so that when I think or say "elm" I mean elm, not beech. When I think or say "Prague" I mean Prague, not Poughkeepsie. And so on.

My words mean what they mean because that's what they mean in English. Sometimes I may misunderstand a word and think that it means something else. But in that case, I'd be wrong, and if I tried to use the word in my peculiar sense, I would miscommunicate and my thoughts would be misguided until I corrected my misunderstanding. Even if Mentalese exists, it wouldn't help until I connected my public speech with these Mentalese denotations, and as Putnam made clear, my Mentalese concepts would have to come into correspondence with things in the world.

As a devoted student of methodological solipsism, I often thought of mental processes giving rise to social behavior. The causal mechanisms, in my mind, were psychological processes; the social world was an emergent property of our individual actions, which themselves result from the operation of psychological mechanisms. I now think that I got the chain of causation exactly backwards. The psychological does not cause the social; rather, social life results in our psychological world.

There is another interesting consequence of this view. We often think of meaning as being tied to grammar, but if the social view is correct, meaning really emanates from the way the grammar is used to signal meaning. In other words, meaning is external to grammar proper. In what follows, I work out how we can create a theory of meaning as the rational use of grammar by social agents.

I'm stating things very strongly here, but think about it. My words get their meaning in more or less the same way that my money gets its value.

It's the untold number of transactions per day that fix the value of the dollars in my pocket, just as it's the social practice of language that fixes the content of my utterances. In fact, I can extend Putnam's hypothesis:

(8) The values of linguistic expressions are fixed by economic and ecological processes.

By *ecological*, I mean that linguistic expressions compete to express meanings in a manner that is analogous to the way that species compete to consume resources and occupy ecological niches. The analogy with economics is close, but not exact. For example, are my words subject to inflation? Probably not in the same sense as money, but money plays a different social role than words do, even if there is an analogy in the way that they receive content.[2]

I've already observed how the content of words is fixed by social exchanges. What about ecological processes? Meaning can be thought of as a kind of resource that linguistic units compete with each other to express. For example, I might tell you that I "text-messaged" someone on my phone, or that I "texted" someone on my phone; both words convey the same meaning. So *text-messaged* and *texted* are in competition to express that meaning. A meaning can be visualized as an ecological niche, and the various linguistic units that can express that meaning as species competing for that niche. See chapter 9 for more discussion.

But there is a further social dimension. In trying to express the meaning, the speaker has to make a choice between *texted* and *text-messaged*. That choice is conditioned by a variety of factors that can be made explicit, for example, the length of the expression or the likelihood that others will understand the meaning of what the speaker said. These choices can be made explicit using tools from game theory.

Part II looks at the tools needed to work through problems involving choice when the outcome of that choice depends on the actions of other agents; this is clearly a kind of social interaction and is the proper domain of game theory. Chapter 9 discussed how competition in a public space can give a social face to the meaning of words.

Further Reading

Language variation is central to linguistics. The concern in this chapter is how languages vary in the way that they encode meanings. There is a vast literature on this, but Talmy (1985) is an excellent resource and a good place to start.

The original dirty faces problem is due to Littlewood (1953). Myerson (1991) gives the problem as the cheating wives problem, and Fagin et al. (1995), in somewhat more innocuous form, as the muddy children problem.

The idea that linguistic meaning is ultimately social is certainly not new. I am quite sure, for example, that I read a philosophical discussion of the relation between the curious monetary system of Yap and the representational theory of meaning, but I have been unable to find the original reference. Nevertheless, the conclusion that linguistic meaning is grounded in social relations is certainly present in Saussure's writing. There is a fascinating relation between the work of Saussure and that of Wittgenstein, although the two philosophers were almost certainly unaware of each other; indeed, Saussure died before the First World War, several years before Wittgenstein rose to prominence. See Harris (1988) for discussion of the parallels between Saussure and Wittgenstein. Mishkin (1989) is a standard textbook on the economics of money, and I recommend Zelizer's (1997) useful discussion as well.

Wittgenstein's *Philosophical Investigations* is particularly important here. I have relied heavily on Kripke's (1982) discussion. Working out exactly what Wittgenstein meant is not an easy matter; happily, a precise exegesis can be left to others, I rely on the intuition that meaning is determined by use. This tradition can also be found in the work of Austin (1975), whose work I return to in chapter 9. Levinson (1983; 2000) works out Austin's argument from the point of view of linguistic pragmatics; see also Searle (1969). Brandom (1994) carefully extends this tradition.

Also important is the work of Putnam (1975; 1981) Searle's (1995) book *The Construction of Social Reality* helped me think about the problem, although I'm sure he wouldn't endorse my response.

The idea that the study of meaning has an ecological component has not been much explored but is one of the great inevitables of linguistics. Some discussion of the idea can be found in Nowak, Plotkin, and Krakauer (1999). Nowak (2006) also includes a discussion of language from the point of view of mathematical biology. Jäger (2007) applies evolutionary game theory to the analysis of the typology of morphological case. The full relation between ecology, game theory, and linguistic meaning remains to be explored.

II GAMES AND TRUTH

4 A Primer on Games

Oh, the Rand Corporation's the boon of the world,
They think all day long for a fee.
They sit and play games about going up in flames;
For counters they use you and me, honey bee,
For counters they use you and me.
—Malvina Reynolds, "The RAND Hymn," 1961

Game theory has a dark reputation, particularly in the humanities. The central thesis of this book is that meaning arises from the rational use of language to signal messages. Game theory itself is concerned with rational decision making where the outcome of a decision depends on the behavior of other agents. My hypothesis is that linguistic meaning arises out of strategic decision-making by rational agents. There is a lot in this hypothesis to arouse suspicion. Are people, after all, ultimately rational decision makers?

For people outside of game theory, the very words *game theory* suggest an icy model of human behavior that views behavior as grounded solely in "rational" self-interest; it seems to counsel greed and self-centered behavior of the sort that led to the economic meltdown of 2008, hardly an optimal outcome by rational decision makers. It's hard not to associate game theory with the Cold War world of the arms race, mutually assured destruction, and credible deterrence. The RAND Corporation, which played a central role in the early development of game theory, is notoriously associated with the policies that influenced early U.S. nuclear strategy, from the arms buildup to the problem of delivering the mail after a nuclear strike. It's hard now, in our post–Cold War world, to reconstruct just how evil the RAND Corporation seemed to many people; the fact that RAND fellows worked on such sinister topics as whether nuclear war is winnable guaranteed a reputation somewhat worse than Halliburton's and marginally better than Satan's.

John von Neumann, one of the central figures in the early history of game theory, argued that the United States should carry out a preemptive strike against the Soviet Union before that country could get the bomb. According to his obituary in *Life* magazine,

> After the Axis had been destroyed, von Neumann urged that the U.S. immediately build even more powerful atomic weapons of their own. It was not an emotional crusade. Von Neumann, like others, had coldly reasoned that the world had grown too small to permit nations to conduct their affairs independently of one another. He held that world government was inevitable—and the sooner the better. But he also believed it could never be established while Soviet Communism dominated half of the globe. A famous von Neumann observation at that time: "With the Russians it is not a question of whether but when." A hard-boiled strategist, he was one of the few scientists to advocate preventive war, and in 1950 he was remarking, "If you say why not bomb them tomorrow, I say why not today? If you say today at 5 o'clock, I say why not 1 o'clock?" (von Neumann 1957, 96)

I think that I can understand von Neumann here. He had left Hungary as a young man to avoid the Nazis, only to have the country seized by the Soviets after the war. The world must have seemed a dark place, ruled by the ruthless, where the best we could hope for was that the worst impulses would be held in check.

The strategy that emerged from that period, mutually assured destruction, was a game of stalemate. Both sides sought to make the cost of nuclear exchange unacceptable to the other side by building enough arms to destroy everything and by persuading the opponent that they had the will and means to retaliate in the event of a first strike.

I grew up haunted by the bomb, like everyone around me. Since my family lived near White Sands Missile Range, my father was convinced that we were a primary target for Soviet thermonuclear missiles and that there would be no surviving a thermonuclear strike. In my boyhood, I started having a recurring nightmare. I was alone in the city, which was completely deserted. I would run through the streets with a growing sense of dread; I felt stalked by a malevolent thing. I turned a corner and saw the mountain that divided the city. Then, behind it was a huge flash. I would always wake up. I had this nightmare well into adulthood. I also often had a sense of nonspecific anxiety, a feeling of imminent catastrophe. It is only recently that feeling—the feeling that something was horribly wrong somewhere—has finally left me. I suspect that many people of my age have shared that feeling of morbid uncertainty. It's just this: in the end, my life—the vivid world of my experience with the attendant rush of sensation, wonderful and bad—is a matter of little consequence in a larger game where I am nothing more than a counter, a cat's-paw in a deadly contest.

The policy of mutually assured destruction (MAD) arises out of a game-theoretic analysis of the problem posed by nuclear weapons. The idea is to achieve an equilibrium state—a strategy that neither party can derive advantage from changing to a different strategy—where the first use of nuclear weapons becomes unthinkable because the opponent will always be able to deliver a mortal counterstrike. It is an uneasy equilibrium, ever haunted by the possibility of error.

Not long ago, I was sitting in my office reading Thomas Schelling's book *The Strategy of Conflict* when a colleague walked in. When he saw what I was reading, he was aghast. "I spent my twenties protesting against this guy," he said. Schelling's work in the early 1960s was devoted to the problem of deterrence. It was based on a notion of punitive retaliation that, while not fatal, would have the effect of making the opponent behave to avoid the retaliation:

> I have known since I was a child that bees can sting and that when they sting they die and that nevertheless they sting. Unable to explain to a bee that its stinging would merely hurt me but would kill it, I have behaved with great respect toward bees. Scores must have lived, because of my anticipation, for every one that died stinging me. (Schelling 2006, 22)

War itself was a kind of bargaining process that used the capacity to hurt to achieve some end.

In mid-1964 National Security Advisor McGeorge Bundy sent a memo to President Johnson that said, "An integrated political-military plan for graduated action against North Vietnam is being prepared under John McNaughton at Defense." McNaughton was a colleague of Schelling at Harvard; the two had consulted about McNaughton's position with the government. In fact, Schelling had proposed McNaughton for his position in the Defense Department. Bundy continued, "The theory of this plan is that we should strike to hurt but not to destroy, and strike for the purpose of changing the North Vietnamese decision on intervention in the south." In a later note that echoes Schelling's thinking, Bundy wrote, "A pound of threat is worth an ounce of action—as long as we are not bluffing."

In an interview reported by Kaplan (1984), Schelling discussed how McNaughton had come to see him for advice. McNaughton explained that the administration wanted to escalate the conflict so that the North Vietnamese would be intimidated. Air power was the obvious answer, but how should escalation be conducted? Schelling had no specific advice except that the bombing campaign shouldn't last more than three weeks; if the bombing hadn't succeeded by then, it would never succeed. On March 2, 1965, the bombing campaign, Operation Rolling Thunder, commenced. It had no impact on the behavior of North Vietnam or the Viet Cong. In December 1967 the Department of Defense announced that the United States had dropped 864,000 tons of bombs on North Vietnam during Rolling Thunder (compared to 503,000 tons of bombs dropped in the Pacific theater during World War II). Estimates of civilian deaths due to Rolling Thunder range from 52,000 to 182,000. The bombing campaign lasted until November 1968. Apparently, they didn't follow Schelling's advice to limit the bombing.

Schelling had no direct responsibility for U.S. policy during that period, but his ideas were influential. Robert McNamara, who was then Secretary of Defense, wrote,

> Between the lines of the statement of the overall objectives of the two-phase bombing program one finds many of the principles espoused by U.S. civilian strategists such as Thomas Schelling.... [His] formulation is elegant, clear, coherent, and—as events later proved—wrong in every important respect. There would be no sign of yielding from Hanoi. Its will was never broken, or even bent....
>
> A story circulated at Harvard during the 1960s that a missed opportunity had occurred when Harvard failed to offer a scholarship to Ho Chi Minh, in order that he might have the opportunity to study with professor Schelling. If he had, according to the Cambridge pundits, he would have known that Washington was trying to send him a *signal* via the bombings. As it was, Ho and his colleagues, in their ignorance, thought the United States was trying to destroy their country. (McNamara et al. 1999, 169–170)

Schelling's interest in signaling, by the way, makes his work particularly relevant to linguistics.

Even parlor games developed a sinister edge. The mathematician and Nobel laureate John Nash was one of the designers of a game called Fuck Your Buddy, marketed as So Long Sucker. The game involved taking opponents' chips as prisoners (while executing one of the chips). The last player alive won. Players were allowed to enter into nonbinding agreements. As the authors themselves describe it,

> This parlor game has little structure and depends almost completely on the bargaining ability and the persuasiveness of the players. In order to win, it is necessary to enter into a series of temporary unenforceable conditions. This, however, is usually not sufficient; at some point it may be to the advantage of a player to renege on his agreement. The four authors still occasionally talk to each other.

This game was invented in 1950 by Messrs. M. Hausner, J. Nash, L. Shapley, and M. Shubik. The aim was to produce an interesting, social game in which coalitions are both profitable and unstable. Technically, it is an essential four-person, no-side-payment game, in extensive form, with perfect information and no chance moves after the first. It has been played extensively in gatherings of different sorts, provoking a wide variety of reactions. The authors will welcome further reactions and comments. (Hausner et al. 1964, 359)

If we look at more recent history, we see a bizarre world of "rational" risk taking with a bestiary of financial instruments, such as credit default swaps and collateralized debt obligations, that seem to defy rational behavior. Would a rational agent, one who was informed and understood these things, risk investing in them? It's far from clear, yet there was an army of highly paid financial analysts who developed sophisticated mathematical models designed to demonstrate the profitability of these instruments. I use *designed* deliberately, since the conclusion they reached seemed to be exactly the one they wanted to reach.

It must seem contrary to logic that I would insist on the communitarian nature of language while relying on tools that emphasize the value of maximizing individual utility. I am aware of game theory's sinister reputation and its rather dark history. Nevertheless, recent work in game theory is illuminating the origins of cooperative behavior, reciprocity, and altruism; these topics are at the heart of the evolution of language. The view game theory takes of individual choice is not only consistent with the social view of language but actually aids understanding of how language works.

Game theory offers the best set of tools for investigating language, both from the point of view of meaning and from the evolutionary perspective. If game theory has had sinister associations in the past, it still suggests how cooperation can evolve.[1] Game theory can deepen our understanding of morality and meaning. This chapter and the next present a normative theory of decision making rather than a descriptive one; a normative theory dictates how an optimally rational agent ought to behave. Of course, this kind of theory describes no actual player. Real agents have real limitations and so can only approximate the rationality prescribed by the normative theory.

Game theory is a branch of mathematics concerned with rational choice where the outcome of the choice depends on the choices of other rational agents. It is a normative theory in the sense that it suggests the choices a rational agent should make in order to optimize his outcome.

Now, think, for a moment, about language and language use. When I say something, I choose what I say with an eye toward its effect on you; that is, I choose what I say *strategically*. When you hear what I say, you interpret it with an eye toward working out what I could have meant. In other words, your interpretation of what I say is also strategic. "What could he possibly have meant," you ask yourself, "by saying that?"

When I ask, "Could you pass the salt?" you wonder, "Why did he ask me that? He knows that I'm perfectly capable of passing the salt, so it can't be a real question." So you don't answer with a yes or no. Instead, being a rational agent, you work out that I must have been making a request; you answer my "question" by handing me the salt shaker. Of course, the example of the salt shaker involves a highly conventionalized ritual, a game everyone knows. I mention it simply for its obviousness. It's far less clear how such a conventionalized exchange involves rationality; I turn to that problem later. For now, I claim that our daily speech exchanges—indeed, our ordinary day-to-day interactions in general—are tactical interchanges where each utterance is a move in a broader game. We plan our moves according to principles of rational behavior; in short, much of our linguistic behavior rests on a foundation of rationality.

Now, if the claim that linguistic behavior is grounded in rationality is to have any substance, we have to determine what rational choice means. We all know that our actions have consequences. Whatever I choose, I must be prepared to live with the consequences. So the first element of a theory of rational behavior should be a collection of actions that an agent is capable of performing. Next, each action should be associated with a consequence of that action. Once actions have been associated with consequences, we then have to say how the consequences stack up in terms of our preferences. Figure 4.1 gives a bit of mathematics to summarize this.

Elements of a theory of rational choice:

- A set A of *actions* from which the decision-maker selects
- A set C of *consequences* to these actions
- A *consequence function* $g : A \to C$ which associates consequences with actions
- A complete, transitive, reflexive *preference relation* $\succsim$ on C

Figure 4.1
Actions, Consequences, and Preferences (Adapted from Osborne and Rubinstein 1994)

Notice that a rational agent's preferences should be ordered by a complete, transitive, reflexive relation. Completeness means that given any two outcomes, *a* and *b*, the agent prefers *a* to *b*, likes both *a* and *b* equally, or prefers *b* to *a*. None of the outcomes can be left out of the preference relation. Transitivity means that if the agent prefers *a* to *b* and prefers *b* to *c*, then the agent prefers *a* to *c*. For example, I might prefer reading the *New York Times* to reading the *Wall Street Journal*. Furthermore, I prefer reading the *Wall Street Journal* to watching Fox News. If I am a rational agent, then it must be the case that, given my already stated preferences, I prefer reading the *New York Times* to watching Fox News. Finally, reflexivity requires that an agent like *a* at least as much as itself. Thus, the preference relation resembles "greater than or equal to," since any number is greater than or equal to itself. Compare this with an nonreflexive relation like "taller than"; John might be taller than Bill, but he certainly cannot be taller than himself.

The Cake Game

To make this discussion more concrete, I consider an actual game. Many parents teach their children fairness by playing a cake-cutting game. In this game, two children are allowed to share a cake by having one child, the slicer, slice the cake into two pieces and having the other child, the selector, select which piece she wants. This is a good example of a game, since each agent's outcome depends on the other agent's choices; both children need to think strategically in order to get the best outcome.

Suppose the slicer has the following possible actions:

(1) a. Cut the cake into two equal slices.
    b. Cut the cake into a large slice and a small slice.

The slicer's actions result in three possible sizes for the cake slices: small, medium, and large. This gives the selector three possible actions:

(2) a. Choose the large slice.
    b. Choose the small slice.
    c. Choose the medium slice.

Notice how the outcomes for the two players are contingent on their individual actions. If the slicer chooses to slice the cake unequally, then the selector can choose between the large slice and the small slice; if the slicer elects to divide the cake into two equal slices, then the selector can only choose a medium slice.

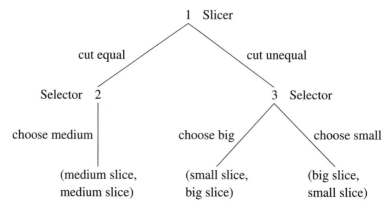

Figure 4.2
Cake Game

Now we turn to the preference relation. In my experience, children generally like cake, so I think I can comfortably assert the following for both children:

(3) a. Big slice ≿ Medium slice.
    b. Medium slice ≿ Small slice.

If the children are rational, they both prefer a big slice to a small slice.

Figure 4.2 shows a simple version of the cake game as a tree. The nodes of the tree are numbered, except for the leaf nodes. I generally refer to the numbered nodes as information states. The top node, information state 1, is the root and represents the choices available to the slicer. He can choose either to cut the cake into equal pieces or unequal pieces. These choices are represented by the branches coming out of the root node; these branches are labeled by the slicer's choices. The next level of the tree represents the selector's choices. If the slicer chose to cut equally (information state 2), the selector has no real choice: there are two medium pieces, and she is indifferent as to which one she takes, since they're the same. Otherwise, the slicer has made unequal cuts (information state 3). The selector can either choose the big slice or the small slice.

The leaf nodes of the tree show the outcomes for the two players. These outcomes are displayed as ordered pairs, with the slicer's outcome as the first element and the selector's outcome as the second element. If the slicer elected to cut two equal pieces, both he and the selector wind up with medium slices. But suppose he cut the pieces unequally. Then, if the selector chooses the big piece, the slicer winds up with the small piece, and if the selector chooses the small piece, the slicer winds up with the big piece.

A Primer on Games

At this point, there is an intuitive result for the game. If the slicer cuts the cake into two equal slices, he's guaranteed a medium slice. If he cuts the cake into two unequal pieces, the selector will take the big piece and the slicer will be left with the small piece. The preference relation tells us that the slicer prefers a medium piece to a small piece, so he'd have to be a fool to cut the cake into unequal pieces. The slicer's best strategy is always to cut the cake into equal pieces:

(1, cut equal).

This is an ordered pair consisting of an information state—in this case, state 1—and an action. So if the slicer is in information state 1, he should choose to cut the cake into equal slices.

From the selector's point of view, things are a little more complicated. If she is information state 2, her choice is determined: she must choose a medium slice. Suppose the slicer makes a mistake and cuts unequal pieces. A rational player would be ready to exploit this mistake. She should choose the big piece, since she prefers it to the small piece. The selector should follow the strategy

{(2, choose medium), (3, choose big)}.

Intuitively, neither player has any reason to change his or her strategy. They are both doing the best they can, given the choices that confront them and the expectation that the other player will follow his or her preferences. We are justified in calling these strategy suggestions an *equilibrium strategy*. If either player defects from the strategy, he or she will do worse. If the slicer cuts unequally, he'll get a small piece. If the slicer makes a mistake and slices things unequally, the selector should choose the big slice, which she likes better than the small piece. We can assemble the choices into a *strategy profile* that tells the players what to do in any event:

{(1, cut equal), (2, choose medium), (3, choose big)}.

It's easy to see why some parents use the cake game to teach fair play. The rational outcome is for the players to share the cake equally.

This analysis of the cake game is straightforward; we don't need much else to understand what's going on. However, it will be useful to have a numerical scale to operate with arithmetically. The decision maker's preferences can be encoded in terms of *utility* (see figure 4.3). For the cake game, the following values for the utility function, U, map from consequences to numeric values of the preferences:

> A *utility function*
>
> $U : C \to \mathbb{R}$
>
> defines the preference relation $\succsim$ by the condition that $x \succsim y$ if and only if $U(x) \geq U(y)$. That is, utility is a numeric reflection of preference: an agent strictly prefers x to y just when the utility of x is greater than the utility of y.
>
> A *rational decision maker* chooses an action $a^* \in B$ that is optimal in the sense that $g(a^*) \succsim g(a)$ for all $a \in B$.
>
> Alternatively, the rational decision maker solves the problem
>
> $\max_{a \in B} U(g(a))$.

Figure 4.3
Some Properties of Utility

$$\begin{bmatrix} \text{Big slice} & \mapsto & 3 \\ \text{Medium slice} & \mapsto & 2 \\ \text{Small slice} & \mapsto & 1 \end{bmatrix}$$

Notice that the utility of a big slice, U(big slice), is 3; the utility of a medium slice, U(medium slice), is 2; the utility of a small slice, U(small slice), is 1. This makes it possible to model the preference relation, $\succsim$, with the "greater than or equal to" relation, $\geq$, since the latter relation is a complete, transitive, reflexive relation on $\mathbb{R}$, the real numbers, and U(big slice) $\geq U$(medium slice) $\geq U$(small slice). In short, the utility function can be used to arithmetize preferences.

We should be careful, though, about concluding what utility really means. It is tempting to reify utility, but it is really just a scale for working with preferences mathematically. Sometimes, you can actually quantify the utility of an outcome by, say, working out how much a player would be willing to pay to join in a lottery that has that outcome as a prize. This is hard to do with linguistic applications; I'm not sure how much I would pay to be able to use a pronoun like *he* instead of a definite description like *the author of the book I'm reading*. Nor is it obvious that different players would get the same utility out of a particular outcome. Thus, keep in mind that utility is being used as a scale. It is clear that people have preferences. In general, I would prefer to use a short expression like *he* instead of a longer description. Of course, the preference rankings given to the outcomes must be justified, but as long as a theory of the preferences is coherent, it should be safe to assign to the outcomes utilities that are consistent with the preferences.

A Primer on Games

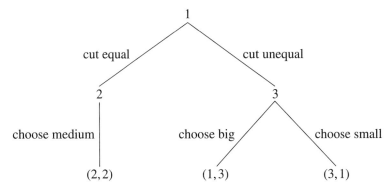

Figure 4.4
Extensive Form of Cake Game

Figure 4.4 shows the full version of the cake game. The game is shown in *extensive normal form*, as a tree where the nodes are information states, the branches are choices, and the leaves are the utilities associated with the outcomes. Further, it is shown as an extensive game of perfect information, which means the following:

1. A set of players has been specified, in this case, Slicer and Selector.
2. There is a set of sequences called *terminal histories*, which have the property that no sequence is a proper subsequence of another. In the cake game, these are
a. (cut equal, choose medium),
b. (cut unequal, choose big),
c. (cut unequal, choose small).
3. There is a player function that specifies which players play when, based on the beginning of terminal history sequences. In this case, this is
a. $P(\emptyset) =$ Slicer,
b. $P(\text{cut equal}) =$ Selector,
c. $P(\text{cut unequal}) =$ Selector.
This is just a fancy way of saying that there is an order of play: first Slicer chooses an action, then Selector chooses a piece.
4. The players' preferences have been specified. This can be seen as a utility assignment for Slicer, u_1, and a utility assignment for Selector, u_2. It's common to define the utility assignments on the terminal histories:
a. $u_1(\text{cut equal, choose medium}) = 2$,
$u_2(\text{cut equal, choose medium}) = 2$.
Both Slicer and Selector assign utility 2 to medium slices; both Slicer and Selector are fairly satisfied with the outcome.

b. u_1(cut unequal, choose big) = 1,
u_2(cut unequal, choose big) = 3.
If Slicer cuts the pieces unequally and Selector takes the big piece, Slicer likes it a little bit and Selector likes it a lot.
c. u_1(cut unequal, choose small) = 3,
u_2(cut unequal, choose small) = 1.
If Slicer cuts the pieces unequally and Selector takes the small piece, Slicer likes it a lot and Selector not so much.

The notation simply serves to emphasize that the game shown in figure 4.4 is a game of *perfect information*. This means that not only do both players know the structure of the game and the utilities associated with the outcomes—these are prerequisites of any rational game playing—but each player knows the other player's choice. Thus, the cake game, as discussed so far, falls into a class of games like checkers, chess, and tic-tac-toe where the player who moves announces her choice to the other player.

Actually, the cake game is so simple that you get the same result if each player knows the other's choice or if the players vote by secret ballot (a game of imperfect information). A rational Slicer will slice the cake equally, and a rational Selector will choose the biggest piece available. Nothing about the information available to the players will alter that outcome.

No other strategy can do better for the two players. The best strategy in the cake game is a *pure strategy*. There is a single optimal course of action; Slicer should never deviate from cutting the cake into equal slices, and if Slicer behaves rationally, Selector will never be able to select anything other than an equal-sized piece. Not all games have pure strategies; some games have a *mixed strategy* where the players select between different choices probabilistically.

Chapter 5 discusses some interesting relations between games of perfect information and logic. Suppose that sentences have truth conditions. (Recall that the truth conditions of a sentence stipulate what the world must be like, given that the sentence is true.) These truth conditions are captured as games between two players who are seeking to establish or refute a sentence. The result is a theory of *verification*—how one would *establish* or *refute* a sentence.

Unlike the cake game, a game like poker is a game of *incomplete information*. In order to play rationally, the players must know the structure of the game, and they know the utilities of the outcomes, particularly since they can see the pot, but they don't know what cards are in the other players' hands. In these cases, there is an information asymmetry; players have privileged information that can affect game play. Games of incomplete information have received a lot of attention over the years. They are interesting because players can do things like bluff, which is really a form of communication. Bridge also has elements of a communication system and so is of some interest for the present purposes.

Games of incomplete information don't necessarily provide a good model for normal linguistic communication. A slightly different model, *games of partial information*, may be better (see part III). In a game of partial information, a player announces a choice, but the choice does not completely clarify the other player's position. Suppose, for example, that I say the word *pen*. Did I mean a pen as in a writing instrument or a pen as in an enclosure for animals? My dictionary gives five different meanings for *pen*; some are nouns (writing instrument, enclosure, a female swan, the internal shell of a squid), and some are verbs (to write something). When I say "pen" you have a number of choices for interpretation but don't know which meaning I intended.

Sequential Games and Backward Induction

In games with sequential moves, like the cake game, there is a straightforward method of finding an equilibrium, called *backward induction* or *rollback*. Many communication games involve a speaker, who first encodes a message and transmits it, and a hearer, who receives the message and selects a content for it. Since this way of thinking about communication is sequential, rollback can be used to discover an equilibrium strategy profile.

As the name suggests, rollback works by starting from the outcome of the game and working backward, considering the best move for obtaining a particular choice, until the root node of the tree is reached. A strategy profile discovered in this way is called a *rollback equilibrium*. Here is an example from Dixit and Skeath (2004).

Suppose there are three players—Emily, Nina, and Talia—who all live in the same neighborhood. Someone is going door-to-door asking people who live in the neighborhood to contribute to a community flower garden. The ultimate size and splendor of the garden depends on how many contribute. Each player is happy to have the garden, happier still if the garden is large and magnificent, but each is reluctant to incur the cost of contributing. If no one contributes, the flower garden will be sparse and miserable, a future repository of litter, as so often happens with vacant lots in the city. Clearly, no one wants that.

For each player, there are four outcomes:

- Outcome A. She does not contribute, but the others do (pleasant garden, and she saves money).
- Outcome B. She contributes, and one or both of the others do (pleasant garden, expensive for her).
- Outcome C. She does not contribute, and only one or neither of the others do (unremarkable garden, but at least she saves money).
- Outcome D. She contributes, and neither of the others do (terrible garden, and expensive for her).

Emily, Nina, and Talia all share the same preference ranking:

Outcome A > Outcome B > Outcome C > Outcome D.

So their possible actions are yes (contribute) or no (do not contribute). Mapping from the outcomes to utilities that reflect the preferences of the players, we can assign utilities as follows:

$$\begin{bmatrix} \text{Outcome A} & \mapsto & 4 \\ \text{Outcome B} & \mapsto & 3 \\ \text{Outcome C} & \mapsto & 2 \\ \text{Outcome D} & \mapsto & 1 \end{bmatrix}$$

Now suppose the person collecting donations for the community garden goes from house to house, first visiting Emily and getting her response, then visiting Nina and getting hers, and finally, visiting Talia and getting hers. Then we have a sequential game. What should each player do to maximize her utility? The results are shown in figure 4.5; the payoffs are listed in the order

(Emily, Nina, Talia).

How should the rollback equilibrium be computed? Start with Talia, since she is the last to move. She is associated with the states E, F, G, and H. At state E, if she says yes, she gets a payoff of 3, but if she says no, she gets a payoff of 4. Since she prefers 4 to 3, she should make the move (E, no).

Next, consider her options at state F. If she says yes, she gets a payoff of 3, but if she says no, she gets a payoff of 2. Since she prefers 3 to 2, she should make the move (F, yes).

Talia also moves at state G. If she says yes, she gets a payoff of 3, but if she says no, she gets a payoff of 2. So at state G she should move (G, yes). At state H, if she says yes, she gets a payoff of 1, but if she says no, she gets a payoff of 2. So she should move (H, no).

A Primer on Games

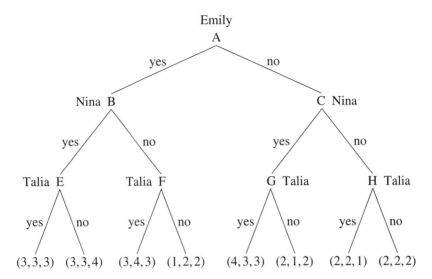

Figure 4.5
Three-Player Game

Talia's best moves are

{(E, no), (F, yes), (G, yes), (H, no)}.

Figure 4.6 shows Talia's best moves circled. The payoffs from her best moves are shown in the nodes one level up, to make it easier to compute Nina's best moves.

Now let's find Nina's best moves. Nina moves at states B and C. In both cases, she knows that however she moves, Talia will go on to make her own best move, so Nina takes that into account. At state B, if Nina says yes, she gets a payoff of 3, but if she says no, she gets a payoff of 4. Clearly, at state B, she should move (B, no). At state C, if she says yes, she gets a payoff of 3, but if she says no, she gets a payoff of 2. So at state C she should move (C, yes).

Nina's best moves are

{(B, no), (C, yes)}.

Figure 4.7 shows Nina's and Talia's best moves circled.

Now let's consider Emily's best move. She moves only at state A. If she says yes, she gets a payoff of 3, but if she says no, she gets a payoff of 4. Clearly, she should move (A, no). Her best move is

{(A, no)}.

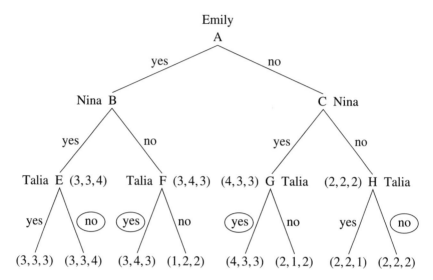

Figure 4.6
Talia's Best Moves

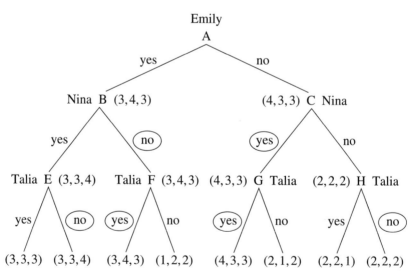

Figure 4.7
Nina's and Talia's Best Moves

A Primer on Games

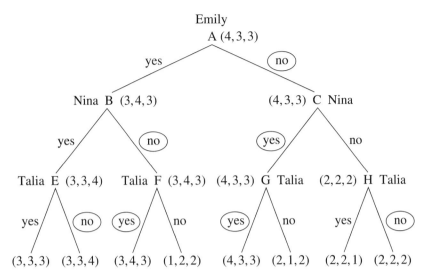

Figure 4.8
Emily's, Nina's, and Talia's Best Moves

The combined best moves for Emily, Nina, and Talia are shown in figure 4.8.

The equilibrium path of play can be found in figure 4.8 by following the branches with circled actions. The entire optimal strategy profile can be found by taking the union of the sets generated at each step:

{(A, no), (B, no), (C, yes), (E, no), (F, yes), (G, yes), (H, no)}.

That is, the optimal strategy profile gives each player's best move at every state. The actual path of play in the rollback equilibrium is found by putting together each player's optimal moves to assemble an actual history. In figure 4.8, this is the sequence found by following the branches with circled actions from the root to the leaf:

(A, no), (C, yes), (G, yes).

Emily clearly has an advantage over Nina and Talia; she has first-mover advantage. The first player to move does not always have the advantage. It has been claimed (but not proved) that in checkers the player who moves second can always force a tie; in that case, there is a second-mover advantage in checkers. Figure 4.9 shows a simple example. Player 2 has a second-mover advantage: if player 1 moves L, then player 2 wins by moving R; if player 1 moves R, then player 2 wins by moving L.

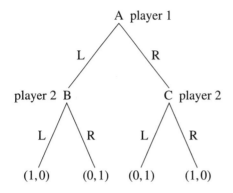

Figure 4.9
Game with Second-Mover Advantage

The Holmes-Moriarty Game

This section discusses a game of incomplete information, a game where one player doesn't know the other's choice when he has to move. In other words, the players are making moves under conditions of uncertainty. Game theory provides players with sound advice about what to do under such circumstances.

This example is from the celebrated book by von Neumann and Morgenstern (1944), which started game theory as an active branch of research. Suppose that Prof. Moriarty is pursuing Sherlock Holmes, no doubt with some evil intent. Naturally, Holmes wants to evade Moriarty. To do so, he has boarded a train bound for Dover, whence he plans to escape to the Continent. As luck would have it, Holmes observes Moriarty on the platform just as the train pulls out. Furthermore, Holmes has every reason to believe that Moriarty saw him and thus knows exactly which train Holmes is on. Even worse, Holmes can safely assume that Moriarty will pursue him in his own private train, which will inevitably overtake Holmes's train before it reaches Dover. The train makes one stop at Canterbury, so Holmes could get off the train there. Of course, Moriarty knows this and could get off his own train as well.

What should Holmes do? Let's set the problem up as a game. Figure 4.10 shows the game in *strategic normal form*. That is, the players' choices are arranged in a matrix, with the payoffs in the cells of the matrix. Holmes is the row player; he has two options: he can get off the train at Dover, or he can get off at Canterbury. Moriarty is the column player; he, too, can descend at Dover or at Canterbury.

A Primer on Games

		Moriarty	
		Dover	Canterbury
Holmes	Dover	−100, 100	50, −50
	Canterbury	0, 0	−100, 100

Figure 4.10
Holmes-Moriarty Game

Notice that the payoffs in each cell sum to zero; Holmes and Moriarty are playing a *zero-sum game*, a strictly competitive game with a clear winner and a clear loser. Traditional logic can be portrayed as a zero-sum game between two agents (see chapter 5).

The cells of the matrix in figure 4.10 can be read as the payoffs assigned to Holmes and Moriarty. In general, the row player gets the first payoff, and the column player gets the second payoff. So if Holmes and Moriarty both descend at Dover, Moriarty catches Holmes, since he got there first and had the chance to prepare a trap. He gets a positive payoff of 100 points. Holmes obviously doesn't like this alternative; he gets a negative payoff of −100 points. The payoffs are the same if both Holmes and Moriarty get off their trains at Canterbury.

Suppose that Holmes makes it to Dover while Moriarty stopped at Canterbury. Holmes is happy, since he can now jump to the Continent where he has more chances to escape the clutches of Moriarty. In this case, Holmes gets a positive payoff of 50, since he has evaded Moriarty and is likely to continue doing so. Moriarty gets a negative payoff of −50. Finally, suppose that Holmes stops at Canterbury while Moriarty continues on to Dover. This is something of a tie; Holmes has, for the moment, evaded Moriarty, but his escape via Dover is unavailable. Moriarty has not yet captured Holmes, but he's well-positioned to do so. Each player gets a payoff of 0.

Holmes, being a brilliant thinker, reasons that the best way to evade Moriarty would be to make him indifferent to the choice between Dover and Canterbury. The idea is that if Moriarty can expect the same payoff if he stops at Canterbury as he can expect if he goes to Dover, then Moriarty becomes indifferent to the choice and can only choose between the two probabilistically. Otherwise, if Holmes adopts a pure strategy, Moriarty could guess it and easily win.

It's somewhat easier to understand the idea with a game like rock-paper-scissors (see figure 4.11), which can be played over and over again.

		Player 2		
		Rock	Paper	Scissors
Player 1	Rock	0, 0	−1, 1	1, −1
	Paper	1, −1	0, 0	−1, 1
	Scissors	−1, 1	1, −1	0, 0

Figure 4.11
Rock-Paper-Scissors Game

The expected utility of a choice $g(a)$ is its utility $U(g(a))$ times its probability, $p_{g(a)}$:

$$EU(g(a)) = p_{g(a)} \times U(g(a))$$

Figure 4.12
Expected Utility

This is a zero-sum game with Rock beating Scissors, Scissors beating Paper, and Paper beating Rock. If a player played another person repeatedly according to a fixed pattern, her opponent would eventually guess her pattern of play and consistently defeat her. The player's best bet is to play randomly, with each option receiving a $\frac{1}{3}$ probability of being played.

This introduces the idea of *expected utility*, which is defined formally in figure 4.12. The expected utility of an outcome is the utility of the outcome times its probability.

Consider the rock-paper-scissors game in figure 4.11. If player 1 is rational, she'll play Paper one-third of the time. If player 2 is rational, he'll play the same random strategy as player 1 is playing. What can player 1 expect when she plays Paper? One-third of the time, her opponent will play Rock, and player 1 will win; one-third of the time, he'll play Paper, and there will be a tie; one-third of the time he'll play Scissors, and player 1 will lose. So player 1's expected utility for playing Paper is

$$\left(\frac{1}{3} \times 1\right) + \left(\frac{1}{3} \times 0\right) + \left(\frac{1}{3} \times -1\right) = \frac{1}{3} + 0 - \frac{1}{3} = 0.$$

If both players are rational, the best player 1 can expect is nothing; rock-paper-scissors is a feckless game unless one plays against a sucker. The

A Primer on Games 83

> Elements of a strategic game:
> - A set of *players*
> - A set of *actions* for each player
> - *Preferences* for each player regarding the expected values of the payoff function over action profiles
>
> A *mixed strategy* of a player in a strategic game is a probability distribution over the player's actions.

Figure 4.13
Mixed Strategies

best way of playing the rock-paper-scissors game is a *mixed strategy*, which is defined in figure 4.13. A player is playing a mixed strategy when she chooses actions probabilistically.

Let's apply the idea of expected utility to Holmes's problem. He wants to make Moriarty indifferent between stopping at Canterbury and continuing on to Dover; in other words, he wants Moriarty's expected utility for picking Canterbury to be the same as his expected utility for picking Dover:

$EU(\text{Canterbury}) = EU(\text{Dover})$.

Now, Holmes will get off at Canterbury with some probability p; since his only other option is to go to Dover, going to Dover has probability $(1 - p)$. So Moriarty's expected utility for getting off at Canterbury is

$[(1 - p) \times -50] + (p \times 100)$,

and his expected utility for descending at Dover is

$[(1 - p) \times 100] + (p \times 0)$.

Holmes needs to figure out the probability, p, that makes the expected utility of picking Canterbury the same as the expected utility of picking Dover; that is, Holmes needs to solve the following equation:

$[(1 - p) \times -50] + (p \times 100) = [(1 - p) \times 100] + (p \times 0)$.

A little algebra reveals that the two sides of the equation are equal when $p = \frac{3}{5}$. That is, Holmes should stop at Canterbury three-fifths of the time and continue on to Dover two-fifths of the time. He can therefore decide by flipping a coin biased toward heads with probability 0.6 and biased toward tails with probability 0.4. Since that kind of coin is hard to come by, particularly when traveling by train, perhaps he should simply take five

> Let
>
> α_i = strategy profile for player i
>
> α^* = strategy profile for all players in a game with the property that no player gets a better payoff by changing his or her action
>
> α^*_{-i} = strategy profile for such a game, excluding player i
>
> Then, a *mixed strategy Nash equilibrium of a strategy game* is defined as follows:
>
> The mixed strategy profile α^* is a (mixed strategy) Nash equilibrium if, for each player i and every mixed strategy α_i of player i, the expected payoff to player i of α^* is at least as large as the expected payoff to player i of $(\alpha_i, \alpha^*_{-i})$ according to a payoff function whose expected value represents player i's preferences. More succinctly, for each player i,
>
> $U_i(\alpha^*) \geq U_i(\alpha_i, \alpha^*_{-i})$
>
> for every mixed strategy α_i of player i, where $U_i(\alpha)$ is player i's expected payoff to the mixed strategy profile α.

Figure 4.14
Mixed Strategy Nash Equilibrium (Adapted from Osborne 2004)

pieces of paper, three with *Canterbury* written on them and two with *Dover* written on them, mix them up, and draw one from his deerstalker cap.

Of course, Holmes is not guaranteed that he will evade Moriarty by playing this strategy. Alas, life has no guarantees. But he can do no better than to play this way. It provides his best chance for escaping Moriarty. Notice that the strategy is a mixed strategy; Holmes should make his choice by probabilistically choosing among several options.

Figure 4.14 defines a *mixed strategy Nash equilibrium*. While the definitions look a bit thorny, the insight that underlies them is surprisingly simple. Suppose that some player, say, Holmes, is trying to decide how to play a game. He observes that all the other players are playing their best, maximizing their expected utilities. Holmes's best strategy under those conditions would be the strategy profile that gives him the best payoff. Once Holmes starts playing this way, then all the players will be maximizing their expected payoffs, and no one has any rational reason to defect to another strategy.

John Nash proved that every game has at least one Nash equilibrium, possibly mixed. Some games might have several Nash equilibria. Then the players will have the problem of selecting from more than one possible Nash equilibrium strategy profiles. We'll turn our attention to this problem in greater detail when we talk about how conventions can arise. For now, we should just keep this fact in mind.

The Holmes-Moriarty game shows how one can work from utilities to probabilities. Because Holmes is aware of Moriarty's utilities, he can work out the probability of stopping at Canterbury. This might seem to reify utility in exactly the way I previously advised against. But utility is still used as a preference scale here. As long as Holmes is aware of the degree to which Moriarty prefers one outcome over the other, he can assign utilities in such a way as to work out a mixed strategy.

"Ideal Free" Ducks and Mixed Strategy Nash Equilibria

In 1982, D.G.C. Harper reported on a series of experiments on the foraging behavior of mallards. The question was whether the ducks would behave in an "ideal free" manner. That is, were the ducks ideal in their assessment of the profitability of various food sources, and were they free to move to the food source of their choice? The experiments were carried out on a flock of thirty-three mallards on a lake in the University Botanic Garden in Cambridge, England, during the winter of 1979–80. Experiment 1 in his paper is a further example of how to compute a mixed strategy Nash equilibrium.

Each day, two experimenters distributed 2-gram bread balls at the lake. They took up positions about 20 meters apart and, at a signal, began to distribute bread balls. Either both threw a bread ball every 5 seconds, or one threw a bread ball every 5 seconds and the other every 10 seconds.

Where should a duck position itself to maximize its access to bread balls? This game has thirty-three players, but it can be simplified by considering a two-player game of one duck versus a flock. If the two experimenters throw bread balls at an equal rate, half the flock should gather around one experimenter and half around the other experimenter. This is exactly what Harper found. More interesting is the unequal-rate case, where one experimenter throws balls at the rate of one every 5 seconds while the other throws balls at the rate of one every 10 seconds.

Call the two experimenters Gupi and Bagha.[2] Assume that all the ducks standing in front of Gupi and Bagha have an equal chance of catching a bread ball. We need to find an equilibrium point for a given duck.

The value of Gupi's position would be a function of the amount of bread Gupi throws over some period of time. Call this $Resource_t(Gupi)$ divided by the average number of ducks stationed before Gupi, $ducks_{Gupi}$:

$$\text{Value}_t(\text{Gupi}) = \frac{\text{Resource}_t(\text{Gupi})}{\text{ducks}_{\text{Gupi}}}.$$

That is, the value of Gupi to a duck over some period of time t is a function of the amount of bread balls he distributes divided by the number of ducks competing for those bread balls.

Of course, the number of ducks in front of Gupi will be a function of the total size of the flock. That is, a fraction of the flock, call it p, will be in front of Gupi:

$$\text{duck}_{\text{Gupi}} = p \times |\text{flock}|,$$

where $|\text{flock}|$ is the actual size of the flock. (In this case, the number is 33). Then

$$\text{Value}_t(\text{Gupi}) = \frac{\text{Resource}_t(\text{Gupi})}{p \times |\text{flock}|}.$$

What do these equations mean? If a duck were alone in front of Gupi during time period t, it would get all the bread, and its utility would be

$\text{Resource}_t(\text{Gupi})$.

However, there are, in fact, some proportion of ducks, $p \times |\text{flock}|$, in front of Gupi. Thus, the probability of a duck's getting a piece of Gupi's bread is

$$\frac{1}{p \times |\text{flock}|}.$$

So the expected utility of sitting in front of Gupi would have to be

$$\text{Resource}_t(\text{Gupi}) \times \frac{1}{p \times |\text{flock}|},$$

or, more clearly, in terms of expected utility,

$$EU(\text{Gupi}) = \frac{\text{Resource}_t(\text{Gupi})}{p \times |\text{flock}|}.$$

Now, like Gupi, Bagha is also distributing bread at some particular rate. Since $p \times |\text{flock}|$ is sitting in front of Gupi, that leaves $(1-p) \times |\text{flock}|$ sitting in front of Bagha. Thus, any one duck's expected utility for sitting in front of Bagha must be

$$EU(\text{Bagha}) = \frac{\text{Resource}_t(\text{Bagha})}{(1-p) \times |\text{flock}|}.$$

The Nash equilibrium of the game would set the expected utility of sitting in front of Gupi to be equal to the expected utility of sitting in front of Bagha, so

$$\frac{\text{Resource}_t(\text{Gupi})}{p \times |\text{flock}|} = \frac{\text{Resource}_t(\text{Bagha})}{(1-p) \times |\text{flock}|}.$$

We can solve for p, the only unknown in the equation:

$$p = \frac{\text{Resource}_t(\text{Gupi})}{\text{Resource}_t(\text{Gupi}) + \text{Resource}_t(\text{Bagha})}.$$

Suppose Gupi is distributing a 2-gram bread ball every 5 seconds (24 grams per minute), and Bagha is distributing a 2-gram bread ball every 10 seconds (12 grams per minute). We can solve for p:

$$\frac{24}{24+12} = \frac{24}{36} = \frac{2}{3}.$$

That is, two-thirds of the ducks should congregate in front of the more generous Gupi, and one-third should congregate in front of Bagha. Out of a flock of thirty three ducks, twenty-two should gather in front of Gupi and eleven in front of Bagha. This is exactly what Harper found. In addition, the ducks were able to solve the problem in about 60 seconds.

Notice that each duck is in an equilibrium state; no one duck can increase its expected payoff by changing strategies, except through kleptoparasitism, stealing from other ducks. (Harper studied the effects of this in some of his other experiments and noted that the ducks do not in fact get equal payoffs.)

Furthermore, each individual duck seemed to divide its time between the two food sources in proportion to the expected utility of that source. That is, each duck spent two-thirds of its time at the more profitable food source and one-third of its time at the less profitable source. Thus, the flock behavior—two-thirds of the flock before Gupi and one-third before Bagha—is a direct consequence of the individual decisions of the ducks. The macrobehavior of the group is the result of the microbehavior of the ducks.

The relation between individual decisions and macroscopic structure has important consequences for the study of language. The patterns found in a community of speakers are the result of individual decisions. In the case of language, however, individuals attempt to coordinate their behavior with community norms. Chapters 6 and 9 discuss linguistic coordination more thoroughly.

Mixed Strategy Nash Equilibria and Language Variation

Why should someone interested in linguistic behavior, and in particular in meaning, be interested in something like mixed strategies? Notice that a player playing a mixed strategy will, to an external observer, behave probabilistically. Sometimes he'll do one thing, and at other times he'll do something else. His behavior will by necessity be described using some kind of probability function. Sociolinguistics offers many examples of variable linguistic behavior that we might try to account for by using a game analysis where the optimal strategy is a mixed one. Variation is part of the fundamental data of linguistics; any linguistic theory that ignores linguistic variability cannot account for why language looks the way it does. In particular, language typology—the way linguistic features are grouped across languages—might be analyzed using evolutionary game theory and some ideas from mathematical ecology.

To take a concrete example, it's not uncommon in English to "drop a g" at the end of some words (actually, a velar nasal becomes an alveolar nasal):

walking $\mapsto$ walkin',

nothing $\mapsto$ nothin'.

Everyone does it, but no one does it all the time.

On the face of it, then, speakers have to make a choice between producing a full form like *walking* and a reduced form like *walkin'*. Could this variable behavior be the result of a mixed strategy?

Let's take an example from the 2008 election. Some commentators suggested that various candidates changed their speaking style, suppressing or accentuating various features of their speech, in an effort to get voters to identify with them. In particular, some people noted that the Republican vice presidential nominee, Sarah Palin, adjusted her speech to suit the occasion. In "The Real Thing," a commentary for the radio show *Fresh Air* (October 14, 2008), Geoffrey Nunberg said,[3]

> So, like Bill Clinton, Palin can signal authenticity simply by refashioning her original accent, rather than acquiring a new one. You can actually hear how this developed if you pull up the YouTube video of Palin as a 24-year-old Anchorage sportscaster fresh from her broadcasting classes in college. She wasn't in control of her accent back then: she scattered the desk with dropped g's: "Purdue was killin' Michigan"; "Look what they're doin' to Chicago."
>
> It's strikingly different from the way she talks now in her public appearances, not just because she's much more poised, but because she's learned how to work it. When she talks about policy, her g's are decorously in place—she never says "reducin' taxes" or "cuttin' spendin'."

But the *g*'s disappear when she speaks on behalf of ordinary Americans—"Americans are cravin' something different" or "People...are hurtin' 'cause the economy is hurtin'." It's of a piece with the *you betchas, doggones* and the other effusions that are meant to signal spontaneous candor.

Now there are clearly a lot of people who find this engaging, but I can't imagine that anybody really supposes it's artless. What it is a stone-washed impersonation of a Mat-Su Valley girl. I wouldn't be surprised if Palin and her friends perfected this way back in high school. There's no group that's so unselfconscious that its members don't get a kick out of parodying their own speech: most Brooklynites do a very creditable Brooklyn, and every Valley girl can do a dead-on Valley girl. And with all credit to Tina Fey, she wouldn't be so brilliant at doing Sarah Palin if Sarah Palin weren't so good at doing herself.

In other words, Palin seemed to be dropping *g*'s *strategically*. Such behavior is exactly the kind of thing we would expect to model with a game. Notice, though, that Palin's choice was conditioned by context. In a policy discussion, her speech was more formal, and she tended to keep the velar nasal (she didn't drop the *g*). But when she wanted some audiences to identify with her, she was more likely to drop the *g*. Her payoff was, in the first case, more respect and, in the second case, more empathy.

So, in the two contexts, the likelihood of the behavior was different. I do the same thing. In a formal setting I'll be more careful about keeping the velar nasal (*singing*)—although probably not all the time—while in an informal setting I'll substitute the alveolar (*singin'*) with a higher probability. The probabilities are *conditioned* by the context, which can be accounted for in terms of the speaker's goals. This is strategic behavior; interestingly, the probability of dropping the velar nasal never goes to 1 or 0, although the likelihood of doing so is clearly conditioned by the context. Of course, we might wonder whether *mixed* strategies are the right way to account for this kind of strategic behavior.

If I'm on the right track here, then the tools provided by game theory can provide an interesting model of meaning and strategic behavior as well as language variation.

Coordination Games

Coordination games are an interesting class of games, where the players' behavior resembles linguistic behavior, in these games the players must coordinate their behavior. On a rudimentary level of analysis, it's clear that any sign system involves coordination of signs with content. For example, a call system for a species of animals might involve a sender, who emits a call, and a receiver, who associates the call with some appropriate behavioral response.

	Receiver			
	Content 1	Content 2	Content 3	Content 4
Message 1	1, 1	0, 0	0, 0	0, 0
Message 2	0, 0	1, 1	0, 0	0, 0
Message 3	0, 0	0, 0	1, 1	0, 0
Message 4	0, 0	0, 0	0, 0	1, 1

Sender (rows)

Figure 4.15
A Simple Call System

	Clem	
	Beans	Spam
Beans	2, 1	0, 0
Spam	0, 0	1, 2

Abner (rows)

Figure 4.16
Hobo Dinner Game

Figure 4.15 shows an example of such a game. The game is quite simple: each message is associated with a single content. There is a positive payoff for associating the right content with a message but no penalty for making a mistake. The call system in figure 4.15 is not much like a human language, where signs are often ambiguous and there can be real penalties for misunderstanding. Nevertheless, language is a clear example of coordinated behavior, so it's worth considering a coordination games in some detail.

Suppose there are two hobos, Abner and Clem. They decide to have dinner together if and only if they can agree on what to have. Abner has a can of Spam, which he would prefer to keep and eat by himself the next evening. He wants to eat whatever Clem has on offer but will share his can of Spam if he has to. Clem has a can of beans, which he would prefer to save for tomorrow, so he would prefer to eat Abner's Spam. Clem will, however, share his beans if that's the group decision. Both Abner and Clem would prefer to eat together and have decided to forgo dinner if they can't decide. Let's call this game the hobo dinner game.[4]

Figure 4.16 shows the strategic normal form of the hobo dinner game. It should be clear that Abner and Clem have two equilibrium choices:

(beans, beans) and (Spam, Spam). Neither player would have any reason to defect from one choice or the other provided that the other player makes the same choice. Abner should choose Spam if Clem chooses Spam, and vice versa. The problem is that they need to coordinate their choices, and their preferences don't coincide exactly.

One way they could reason about the game would be on analogy with Holmes's reasoning in the Holmes-Moriarty game (see figure 4.10). Suppose Clem assigns probability q to beans. Abner's expected utility for choosing beans is

$$2q + [0 \times (1 - q)] = 2q,$$

while Abner's expected utility for choosing Spam is

$$(0 \times q) + 1 \times (1 - q) = 1 - q.$$

So if

$$2q > 1 - q,$$

then Abner should always pick beans; his probability for picking beans is 1. If

$$2q < 1 - q,$$

then Abner should always pick Spam; his probability for picking beans is 0. Suppose, however, that $q = \frac{1}{3}$:

$$2q = 1 - q.$$

In that case, any probability Abner assigns to picking beans will work. Of course, the point at which he is indifferent to choosing beans or choosing Spam is when he selects beans with probability $\frac{2}{3}$.

Similar, reasoning applies to Clem. The point at which Clem becomes indifferent to choosing beans or choosing Spam is when he assign probability $\frac{1}{3}$ to choosing beans.

Combining the two viewpoints, we discover that the hobo dinner game has three mixed strategy Nash equilibria. Let p be the probability that Abner picks beans and q be the probability that Clem picks beans. Then the three Nash equilibria, (p, q), for the hobo dinner game are $(0, 0)$, $(\frac{2}{3}, \frac{1}{3})$, and $(1, 1)$. That is, the hobos should both always pick Spam, both always pick beans, or Abner should pick beans $\frac{2}{3}$ of the time and Spam $\frac{1}{3}$ of the time; Clem should pick beans $\frac{1}{3}$ of the time and Spam $\frac{2}{3}$ of the time.

Notice that this last strategy profile means that Abner and Clem agree on dinner only part of the time, but it's also fair. When they do have dinner together, sometimes Abner gets his way and sometimes Clem gets his way. Of course, one could say that the best course for Abner and Clem would be to negotiate.

Coordination problems are not far removed from everyday use of language. Suppose, for example, that we're talking, and I want to refer to some particular object, say, a particular fountain pen of mine. It happens that I have two fountain pens, so simply saying "my fountain pen" may not work. I need to find an expression that will allow you to coordinate with me; I want you to pick out the fountain pen I have in mind, and you want to know which object I'm talking about. We both want to accomplish some task. As the speaker, I try to send a signal that I think will allow you to understand what I mean. As the hearer, you're trying to determine what I intend by my signal.

Now, imagine that you're in my study and I call to you:

(4) Could you bring me my fountain pen on the desk?

Suppose there is a fountain pen on my desk. Then we're in luck. We've coordinated our behavior, and you bring me the pen you find on my desk. But suppose I got it wrong. I've forgotten that I actually moved the pen. You look around and see that my favorite fountain pen—the monogrammed tortoiseshell one my wife gave me—is on the bookshelf next to my desk. Do you bring that pen to me, or not? If you bring it to me and it's the one I meant, we're both happy. You might decide the expected payoff is worth the risk of being wrong. Or you might be averse to the risk of being wrong. You might then ask for clarification. We both have to do more work, so the payoff is lower but guaranteed. Suppose there is no fountain pen but a rollerball pen. Do you bring me the rollerball pen, taking the risk that this is the coordinating behavior I was looking for?

Although the example is simple, it raises a number of issues. Language is rife with ambiguity. Whenever I produce an ambiguous expression, my audience has to choose from among a set of possible meanings that I could have intended. This is a coordination problem.

Further, the coordination problem entails a certain measure of risk; you might misunderstand what I say. You have to decide whether to take a *payoff-dominant* or a *risk-dominant* approach to interpreting my request. The game in figure 4.17 has two Nash equilibria: (cooperate, cooperate) and (defect, defect). We both do better if we cooperate because each will get the highest payoff. Choosing to cooperate would be payoff-dominant; this Nash equilibrium is sometimes called *Pareto-dominant*. A Nash equilibrium is Pareto-dominant if it pays off at least as much as any other Nash equilibrium.

	Cooperate	Defect
Cooperate	5, 5	0, 2
Defect	2, 0	2, 2

Figure 4.17
Coordination Game

The proper definition of risk dominance is a bit more involved. For the moment, just observe that if I play Defect in the game in figure 4.17, I'm guaranteed a payoff of 2 and I risk nothing. If I play Cooperate, I might get a payoff of 5, but I risk a payoff of 0. If I don't trust you, I might be afraid that you'll defect and I'll get nothing, so I might prefer to defect myself and give up the possibility of the higher payoff.

My hypothesis is that linguistic conventions help us to make payoff-dominant choices; I can trust you to make the right choice because we share a convention. To the degree that there's no general convention, we tend to make risk-dominant choices. The following sections present some games involving conventions and cooperation. Linguistic behavior is deeply social; it involves social coordination almost inevitably. We would expect, then, that coordination games would provide a laboratory for thinking about linguistic signals and their evolution. Chapter 9 discusses the linguistic issues associated with conventionality.

The Prisoner's Dilemma

Perhaps the best-known game studied by game theorists during the Cold War was *Prisoner's Dilemma*. Two suspects in a major crime are brought in for questioning by the police and held in separate cells. The police tell them that they have enough evidence to convict both of them of a lesser crime, which guarantees one year in prison. If they both stay silent, they are both sure to face the one-year prison sentence, but they will be assured that they won't have the reputation of being a rat. If, however, one rats out the other, the informer will be set free and the other will face the full penalty for the major crime. If they both inform on each other, then both will get three-year prison sentences.

The strategic form of the game is shown in figure 4.18. The payoffs have the property that for both players, the preferences arrange themselves so that the informer gets a big payoff—the temptation payoff—and the other player gets nothing—the sucker payoff—if the informer speaks out while the other stays silent. Notice that I've ranked the

	Player 2	
	Inform	Silence
Player 1 Inform	1, 1	3, 0
Player 1 Silence	0, 3	2, 2

Figure 4.18
Prisoner's Dilemma Game

preferences from 0 for a full prison term to 3 for freedom. If both cooperate and stay silent, they both get a moderately good payoff. If they both inform, they both get a payoff that is only slightly better than the sucker payoff:

u_1(inform, silence) > u_1(silence, silence) > u_1(inform, inform)

> u_1(silence, inform),

u_2(silence, inform) > u_2(silence, silence) > u_2(inform, inform)

> u_2(inform, silence).

The theoretical basis for prisoner's dilemma game was discovered in 1950 by two scientists at the RAND Corporation, Merrill Flood and Melvin Dresher,[5] and it excited enormous interest. The immediate problem is that the game seems to have an equilibrium state that fails to yield the best payoff. There is an obvious appeal to playing Silence, but any gain from playing it is undercut by the temptation to take a free ride and play Inform. Thus, Silence is dominated by Inform; only a fool plays a dominated strategy, so (inform, inform) is the equilibrium state, but its payoff is only slightly better than the sucker payoff.

Let's first confirm that (inform, inform) is indeed the equilibrium state; my own intuition when I first read about the prisoner's dilemma game was that it should be (silence, silence). But consider the state (silence, silence). Player 1 should be drawn to the temptation payoff he gets by defecting to Inform. This is also true of player 2, so we have symmetrical logic; we need only consider the perspective of player 1 to make the point. We see immediately that the state (silence, silence) can't be an equilibrium state. Now consider the state (inform, silence), which player 1 is drawn to. This can't be an equilibrium state either because player 2 will do better by defecting to Inform.

Now consider the true equilibrium state, (inform, inform). Neither player can do better by changing his or her own choice. Of course,

A Primer on Games

	Straight	Veer
Straight	−1, −1	1, 0
Veer	0, 1	0, 0

Figure 4.19
Sidewalk Game (I)

either player would do better if the other player changed his own strategy, but that's not going to happen because each player is playing according to his own interest. At first glance, we see a rather dark conclusion: if a player behaves rationally, according to his own interest, he does rather poorly. If he behaves communally, in a way that maximizes everyone's payoff, he is simply laying his neck on the chopping block so that the other player can win. As Hobbes wrote in *Leviathan* (1651) about life during war,

> Whatsoever therefore is consequent to a time of war, where every man is enemy to every man, the same is consequent to the time wherein men live without other security than what their own strength and their own invention shall furnish them withal. In such condition there is no place for industry, because the fruit thereof is uncertain: and consequently no culture of the earth; no navigation, nor use of the commodities that may be imported by sea; no commodious building; no instruments of moving and removing such things as require much force; no knowledge of the face of the earth; no account of time; no arts; no letters; no society; and which is worst of all, continual fear, and danger of violent death; and the life of man, solitary, poor, nasty, brutish, and short.

A bleak picture indeed! Because winning big in the prisoner's dilemma game means cutting the throat of a cooperative person, it implies that we can do best by anticipating the other player's bad behavior. "Fuck Your Buddy" seems to provide the dark beacon of our social lives.

What's worse, the Prisoner's Dilemma seems to be a model for a variety of social interactions. These are all cases where people are seemingly forced by unkind circumstances to accept a meager payoff. Consider the game shown in figure 4.19. It's not quite the same as the prisoner's dilemma game, but it has the same property of having an unattractive equilibrium.

The sidewalk game (or *Chicken*) is what happens when two people approach each other in a collision course on the sidewalk. A small *pas de deux* often ensues where one person tries to avoid the other by veering in one direction or another. I've worked out the preferences based on the idea that each person would prefer to continue in a straight line if possible

		Car 2	
		Stop	Go
Car 1	Stop	0, 0	0, 1
	Go	1, 0	−1, −1

Figure 4.20
Stop Sign Game

but doesn't want to collide. The other possibility is to veer in one direction or the other. Notice that the strategy (straight, straight) is not an equilibrium, since each player gains by defecting to Veer. Thus, it would seem that both players should play Veer, since (veer, veer) is the equilibrium. Of course, this doesn't happen that often on a real sidewalk.

Consider another simple game (figure 4.20). Two cars arrive simultaneously at a stop sign; both would prefer to go immediately but will wait to avoid a crash. Once again, the equilibrium state, (stop, stop), yields no one's optimal payoff; in fact, it seems that it leaves drivers trapped eternally at the intersection.[6]

This impasse is solved because there is a convention about how to behave at stop signs. The first driver to arrive at a stop sign gets to proceed first; if two (or more) drivers stop simultaneously at a controlled intersection, the one on the left has to yield. This convention maintains a smooth flow of traffic and minimizes impasses.

The prisoner's dilemma game is often taken as offering an interesting way to think about conventions. What happens if one plays Prisoner's Dilemma over and over with the same person? If both play Inform, then each gets 1 point per round. If both trust each other to play Silence, then both get 2 points per round. Intuitively, neither player has reason to defect; it seems as though, on repeated plays, the equilibrium state changes. I am skeptical about using the prisoner's dilemma game as a model for conventions, particularly linguistic conventions, but understanding the game is useful for thinking about other games that work better as models of social behavior.

Robert Axelrod, a political scientist at the University of Michigan who has worked extensively on the evolution of cooperation, notes that during the trench warfare in World War I, the soldiers spontaneously developed a kind of tit-for-tat strategy. The soldiers were trapped in a static situation, facing the same enemy for extended periods. Both sides began lifting their guns to fire above the other side. When ordered to do so, the troops would fire directly on each other, but between large battles they deliberately avoided hitting each other, provided that the other side did so as well. If one side fired directly on the other side, then the other side would reply by firing directly on the first. This would go on for a round or two, and then the guns would be raised again. This convention arose spontaneously without the need for either side to openly negotiate with the other.

It makes sense that playing the same game repeatedly with a person or group would change the equilibrium state. Suppose I opened a coffee shop and decided to make the coffee out of the cheapest ingredients available—gutter water and pencil shavings, perhaps—and sell it at enormously inflated prices. I might make a vast profit per cup that I sold, but I'm virtually guaranteed not to have any repeat business, and my coffee shop will certainly perish in bankruptcy court. It makes more sense for me to be cooperative and make the best cup of coffee I can, given a small but fair profit per cup, and hope that customers will come back to play the next day.

Axelrod, in fact, staged a contest where computers repeatedly played the prisoner's dilemma game with each other using different strategies that contestants had proposed. The best strategies involved cooperation; in fact, the winning strategy was Tit for Tat (TFT). This strategy says cooperate unless the opponent defects, then defect once, and return to cooperation. Another strategy might be to cooperate unless the opponent defects, then always defect. When either strategy is played, the result is consistent cooperation.[7]

Could we spontaneously "evolve" conventional behavior by allowing agents to play repeated games with each other? We are constantly negotiating our language with each other. I used to say "text-messaging," but the term "texting" took over. How does a population arrive at conventional labels for new technology? Or, more generally, how do we arrive at a common vocabulary? One reason I refer to a dog with the noun *dog* is that everyone around me does the same thing. We all have a stake in sticking with this convention.

One way to study this would be to play a game with a large number of simple computing agents and study the conditions under which this population will settle on a stable strategy. This is basically how linguistic conventions could arise.

Let's return to the sidewalk game. While there is an explicit convention for behavior at stop signs, there is no such convention for negotiating space on the sidewalk. Figure 4.21 shows a more complex version of the

	Veer Left	Straight	Veer Right
Veer Left	1, 1	1, 2	−1, −1
Straight	2, 1	−1, −1	2, 1
Veer Right	−1, −1	1, 2	1, 1

Figure 4.21
Sidewalk Game (II)

sidewalk game, which is still somewhat simplified but more realistic than the version in figure 4.19. If we are facing each other and I choose to go right while you choose to go left, we'll collide; equally, we'll collide if you go right and I go left. This actually happens fairly frequently; people decide to veer but veer in such a way that they have to veer again to avoid a collision. The result is they are trapped in an absurd dance.

In the game in figure 4.21, there are two negotiations that must occur in order for the players to cooperate. They have to decide whether one or both will veer. If both veer, they need to negotiate in which direction to veer. Since they don't have much of an expectation about what the other will choose, they have trouble settling on a good strategy. Old hands at city walking know there is a certain amount of tacit signaling that goes on. Body language serves as a channel of communication that lets a pedestrian know what another plans to do. For the most part, though, pedestrians seem to have adopted the convention that the sidewalk is like a two-lane road; they keep oncoming pedestrians to the left. The madman approach used to be quite effective: simply mumble and look dazed. Other people would actually cross the street to avoid you. The widespread adoption of cell phones seems to have rendered this stratagem less effective.

Cooperation: The Stag Hunt

I tend to have a sunnier disposition than the prisoner's dilemma game, or even experience, would advise. The prisoner's dilemma game arose in a particular historical context, the Cold War, and it reflects the stresses and anxieties of that period. Certainly, the prisoner's dilemma game merits attention; Axelrod's studies of repeated prisoner's dilemma games provide grounds for hope. But is Prisoner's Dilemma a model for human cooperation?

		Hunter 2	
		Deer	Hare
Hunter 1	Deer	2, 2	0, 1
	Hare	1, 0	1, 1

Figure 4.22
Stag Hunt Game

Virtually everything around us offers evidence of human cooperation. If we were truly solitary agents, our lives would indeed be "poor, nasty, brutish, and short." Instead, we have the material and emotional comforts of society. This suggests that the model of Prisoner's Dilemma is not the whole story. Fortunately, there is another simple game that seems to say a great deal about cooperation.

The game is described in *Discourse on Inequality* (1754) by the Enlightenment philosopher Jean Jacques Rousseau; he does not provide a very rosy picture of human behavior:

> In this manner, men may have insensibly acquired some gross ideas of mutual undertakings, and of advantages of fulfilling them: that is, just so far as their present and apparent interest was concerned: for they were perfect strangers to foresight, and were so far from troubling themselves about the distant future, that they hardly thought of the morrow. If a deer was to be taken, every one saw that, in order to succeed, he must abide faithfully by his post: but if a hare happened to come within the reach of any one of them, it is not to be doubted that he pursued it without scruple, and, having seized his prey, cared very little, if by so doing he caused his companions to miss theirs.

In other words, if the hunters cooperate and remain focused, they can catch a deer, which is a high payoff. It's always possible, though, that one (or more) of the hunters will defect and catch a hare. He gets a small payoff with the outcome that the other hunter gets nothing. Rousseau seems to assume that the hunters will immediately go for the small-payoff hare, but I think that many hunters will cooperate for the high-payoff deer.

Figure 4.22 shows the stag hunt game in strategic form. The game has two Nash equilibria: (deer, deer) and (hare, hare). It should be obvious that (deer, deer) is a Nash equilibrium, since it pays both players the most that they can get.

Let's confirm that (hare, hare) is also an equilibrium. Its payoff is mediocre for both players. Notice, though, that neither player can do better

> Given a game of the following form:
>
> | | M | N |
> |---|---|---|
> | M | A, a | B, b |
> | N | C, c | D, d |
>
> Strategy pair (M, M) *Pareto-dominates* (or *payoff-dominates*) (N, N) if $A \geq D$, $a \geq d$, and at least one of the two is a strict inequality: $A > D$ or $a > d$.

Figure 4.23
Pareto Dominance

by unilaterally defecting to another strategy. If hunter 1 plays Hare, then hunter 2 gets nothing if he plays Deer. So, by definition, (hare, hare) is a Nash equilibrium.

Now we're in an interesting position. There are two Nash equilibria, one of which, (deer, deer), pays better than the other. The payoff of the lower-paying strategy, (hare, hare), is a done deal; if a hunter plays Hare, she will get one point no matter what the other does. If a hunter plays Deer, she will have to trust that the other will also play Deer; if the opponent doesn't play Deer as expected, the trusting hunter will be left with nothing.

This brings up the problem of *equilibrium selection*. In the case of the stag hunt game, there are two ways of thinking about this. First, a hunter could decide to trust her opponent—meaning that she has faith in him as a rational agent who understands the game and is seeking to maximize his outcome—and select the highest-paying Nash equilibrium. Recall the definition of a Pareto-dominant Nash equilibrium. Figure 4.23 gives a slightly different definition.[8]

Second, a hunter could decide not to trust her opponent to behave rationally; perhaps he doesn't understand the game, isn't interested in maximizing his utility, or simply makes mistakes because of "trembling hand perfection" (he is a rational agent, knows what his best strategy is, but his hand trembles), so he makes mistakes. The untrusting hunter is averse to risk and prefers a guaranteed payoff to a higher payoff that entails some risk. Because she risks a smaller payoff when she plays Deer—zero, in fact—she will play Hare. Figure 4.24 outlines a couple of ways of thinking about risk dominance.

> Given a game of the following form:
>
> | | M | N |
> |---|---|---|
> | M | A, a | B, b |
> | N | C, c | D, d |
>
> Strategy pair (N, N) risk-dominates (M, M) if the product of the deviation losses is highest for (N, N), in other words, if the following inequality holds:
>
> $(C - D)(c - d) \geq (B - A)(b - a)$.
>
> If the inequality is strict, then (N, N) strictly risk-dominates (M, M).
>
> The risk factor of an equilibrium can be computed in much the same way as the mixed strategy of a Holmes-Moriarty game.
>
> First compute the expected utility for the row player of playing M:
>
> $EU(g(M)) = pA + (1 - p)B$,
>
> where p is the probability that the other player will play M.
>
> Then compute the expected utility for the row player of playing N:
>
> $EU(g(N)) = pC + (1 - p)D$.
>
> The value of p that makes
>
> $EU(g(M)) = EU(g(N))$
>
> is the risk factor for the equilibrium (M, M), and $(1 - p)$ is the risk factor for the equilibrium (N, N). The row player then selects the equilibrium with the smallest risk factor.

Figure 4.24
Risk Dominance

The stag hunt game is certainly less dramatic than Prisoner's Dilemma, but it provides an interesting model for certain types of social behavior. Consider, for example, the question of charitable contributions. I can make a contribution to a charity, and if enough people contribute, the charity has the means to carry on its work. We all benefit, if only indirectly. On the other hand, I can keep my money. Perhaps I can get a free ride, but if enough people do so, then the charity collapses and whatever work it does won't get done. I've gotten a small gain at a larger social cost.

The stag hunt can also be taken as a moral justification for taxation. If I pay taxes, then government has money available for public works

A stag hunt arms race:

	Disarm	Arm
Disarm	2, 2	0, 1
Arm	1, 0	1, 1

A prisoner's dilemma arms race:

	Disarm	Arm
Disarm	2, 2	−1, 3
Arm	3, −1	1, 1

Figure 4.25
Two Ways of Thinking about Disarmament

projects that are both necessary and improve my life. Thus, by being cooperative and paying my taxes, I get things that I couldn't otherwise get. Suppose I keep my tax money in my pocket; I can free-load on other people's cooperativeness for a while, but if enough people act as I do, then the government can no longer fund public works. The roads I rely on to get food to the market fall into disrepair; social costs rise. I've taken a small personal benefit at the expense of a greater social good. Analogous arguments can be made for aid to developing countries; we could keep the money and let the rest of the world fend for itself—with all the potential costs that come with greater disease and poverty in the developing world—or we can cooperate and participate in foreign aid and improve conditions for everyone.

Figure 4.25 shows two ways of thinking about disarmament. The stag hunt version and the prisoner's dilemma version differ as to how they rank the outcomes. In the stag hunt version, both sides prosper most by disarming jointly. Both would prefer not to arm but get some utility from having arms. If one side unilaterally disarms, it is weak relative to the other side and so gets nothing. Of course, this version has two Nash equilibrium strategies: the Pareto-dominant strategy is (disarm, disarm), and the risk-dominant strategy is (arm, arm). A proponent of this version might argue that while no one wants to take the chance of being weak, the stronger side would be morally barred from annihilating the weaker side, so the cost of weakness is not terrible.

The prisoner's dilemma version of disarmament is somewhat more sinister. Both sides prefer jointly disarming to building arms. If one side disarms while the other builds arms, it is likely to be defeated militarily (or worse, obliterated), so that side gets the sucker payoff, and the side that built arms gets the temptation payoff. There is a single Nash equilibrium, (arm, arm). Neither side wants to be taken as a sucker. A proponent of this version would argue that the stronger side would certainly exploit its advantage, at very least to impose unacceptable conditions on the weaker side.[9]

This illustrates that while game theory can provide a formal framework for thinking about strategic decision making, it's still up to us to justify the assumptions. Much linguistics work is not just devising games that model behavior but justifying the preference rankings.

The stag hunt game has an application to linguistic behavior. Often, we mean more than we say; what we don't say can carry as much meaning as what we do say. An utterance can be taken as carrying implicational content as well as its literal content (see chapter 8). The speaker can always deny that she intended to transmit the implied content, a process known as *implicature cancellation* Should the hearer infer that the speaker intended the implied content, something which can involve risk, or should he simply accept only the literal content? The latter process has a lower payoff: less information is transmitted. Thus, there is a choice between a payoff-dominant equilibrium (literal content plus implied content) and a risk-dominant equilibrium (literal content only). Following is an example.

The other morning was a particularly cold one, so I let my wife sleep in while I got up to start the heat, make coffee, and so on. When she got up, she joked, "You want a lazy wife!" I retorted,

(5) I didn't WANT a lazy wife.

At which point she pretended to get mad at me (at least I hope she was pretending). While the literal content of my utterance was that I didn't want a lazy wife, I implicated (a special way of saying "implied") that I got one anyway. Of course, I didn't say the last part, and I could always deny that that's what I meant—I did and do deny it! The example is really very complex. I wanted my wife to get the implication, but I wanted her to get it *as a joke*.

There is a literal content to what I said in (5), which is simply that I didn't want a lazy wife; in addition, there is the implied content that I got one anyway. Notice that I can deny that I intended to imply that my wife is lazy; this act of denying an implication is called cancellation:

(6) I didn't WANT a lazy wife and, happily, I didn't get one.

		Hearer	
		Literal + Implied	Literal only
Speaker	Literal + Implied	3, 3	1, 1
	Literal only	1, 1	2, 2

Figure 4.26
Implicature Game

The sentence in (6) is not a logical contradiction in the way that (7) is:

(7) The sky is blue and the sky is not blue.

The act of packing an implication along with the literal content of a sentence is called a *conversational implicature*, and it's a very handy device because it offers a cheap way of transmitting more than we actually say (see chapters 7 and 8). It's hard to put the subtleties of conversational implicature into strategic normal form, but figure 4.26 shows a simplified version.

In this game, there are two players: a speaker and a hearer. Although it isn't evident in strategic normal form, there is a temporal aspect to the game. The speaker begins transmitting a message, and the hearer tries to assign a content to the message.

The speaker has two moves available. One move is to transmit the message with the intention that the hearer gets both the literal content and the implied content of the message. In this case, the speaker hopes that the hearer will pick up on the proper *implicature*. The other move is to transmit the message intending only the literal content. For instance, I would have said the sentence *I didn't WANT a lazy wife* intending only to deny the assertion that I wanted a lazy wife.

The hearer also has two moves. She can posit the literal content of the message and pick up on the implicature. If she does this, and the speaker intended the implicature as well as the literal content, then all is well. If she picks up on the implicature, and the speaker didn't intend her to, that's a misunderstanding; she got the literal content, which the speaker no doubt intended, but she concluded more than the speaker wanted.

On the other hand, the hearer might conclude that the speaker intended to communicate only the literal content of the message. Again, if the speaker indeed intended to communicate only the literal content, then all is well. But if the speaker intended that the hearer understand the implicature, then things have gone awry. For instance, I heard the following conversation:

(8) *Man*   Do you have a watch?
   Woman   Yes.
   (*Long pause*)
   Man (*irritated*)   Well?
   Woman   Oh! I left it at home.

Here the man intends that the woman draw the inference that he is requesting her to tell him the time. Instead, she simply takes the literal content of his utterance, the question of whether she has a watch. When she fails to supply the time, he becomes audibly annoyed, at which point she draws the intended inference (presumably) and explains why she can't comply with his request.

I've attached utilities to the outcomes of the game according to the following rules of thumb:

- Transmitting the intended information is better than not.

If the hearer fails to get the intended content, that's a miscommunication.

- Shorter utterances are better than long ones.

If the speaker can transmit more information by implicature (and thus say less), that's desirable. If the hearer can pick up on the intended implicature (thus having less utterance to process while still getting the speaker's intended meaning), so much the better.

One might also argue that miscommunications incur penalties.

The game in figure 4.26 has two Nash equilibria: (literal + implied, literal + implied), which is payoff-dominant, and (literal only, literal only), which is risk-dominant. In fact, it looks like a classic stag hunt game. We might hypothesize that players who are in a position of mutual trust are more likely to play a payoff-dominant strategy, while strangers, who have no reason to make particular assumptions about each other, might be more likely to play a risk-dominant strategy.

Of course, the hearer in the watch exchange could have directly inferred that the speaker wanted to know the time and give it to him. Some exchanges, like *Could you pass the salt?* are highly conventionalized. The more conventionalized an exchange is, the more likely it is that mutual strangers will play a payoff-dominant strategy for the exchange because they can trust each other to know the game and maximize their payoffs. Under what conditions will an exchange become conventionalized, and when will players adopt payoff-dominant strategies? The answers lie in studying applications of game theory to populations as well as evolutionary games.

Evolutionary Games

Classical game theory, which has been considered in this chapter, is concerned with prescribing the choices made by rational agents. Evolutionary game theory is concerned with rational choices by a population; the population need not be made up of rational agents, but the eventual outcome of the system will be an equilibrium state. In other words, evolutionary game theory is concerned with how rational choice can emerge from a population of potentially nonrational agents. *Nonrational* here is used in the special sense of game theory; the individual agents are not necessarily aware of the structure of the game they are playing, need not have all the necessary information, and may not be seeking to maximize their own individual payoffs.

Suppose there is a population of individuals, each of which has a mode of behavior, that is, a strategy that it follows in playing a game. The game can be any behavior that involves the interaction of two individuals. The individuals are randomly paired and play the game. The outcome of the game determines how utility is apportioned to the individuals.

Utility has a very precise interpretation in terms of reproductive fitness. The idea is simple and striking. Two organisms engage in a strategic interaction that results in some apportionment of fitness. Suppose the first organism has the more successful strategy and, as a result, wins the game; then it has the higher reproductive fitness and will have more offspring to which it will pass its successful strategy. The organism with the lower fitness—the loser of the game—will have fewer offspring, and its strategy will become less frequent; eventually it will vanish from the population.

Assume further that mistakes can be made in copying a strategy from a parent to its offspring. In this way, new strategies can enter the population and compete to survive. Eventually, the population should reach a steady state, an equilibrium. This equilibrium should be evolutionarily stable in the sense that no organism can do better by defecting from it. Of course, this is just the intuitive definition of a Nash equilibrium. The idea is summarized in figure 4.27.

Consider a simple model of animal interaction, the hawk-dove game. Suppose two animals from the same species compete for a resource, say, territory. Each animal can be either aggressive or passive. If both animals are aggressive, they fight until one of them is seriously injured. The winner gets the territory, with some value, v; the loser suffers some cost, c. For simplicity, suppose each animal is equally likely to win, so the expected payoff of an encounter between two aggressive animals is $\frac{1}{2}(v - c)$.

A Primer on Games

> An *evolutionarily stable strategy* (ESS) is defined by Maynard Smith (1982, 10) as follows:
>
> "A 'strategy' is a behavioural phenotype; i.e., it is a specification of what an individual will do in any situation in which it may find itself. An ESS is a strategy such that, if all the members of a population adopt it, then no mutant strategy could invade the population under the influence of natural selection."

Figure 4.27
Evolutionary Stability

	Aggressive	Passive
Aggressive	$\frac{1}{2}(v-c), \frac{1}{2}(v-c)$	$v, 0$
Passive	$0, v$	$\frac{1}{2}v, \frac{1}{2}v$

Figure 4.28
Hawk-Dove Game

Now suppose that one animal is aggressive and the other is passive. Then the aggressive animal wins the territory without a fight and gets a payoff v. The passive animal loses the territory but doesn't suffer an injury, so its payoff should just be zero.

Finally, suppose that both animals are passive. Then the territory is allocated to one of them by chance. In this case, the expected utility is $\frac{1}{2}v$. Figure 4.28 shows the matrix for the hawk-dove game.

Notice that in the game in figure 4.28, the strategy (passive, passive) is not a Nash equilibrium, since both players will be tempted to change their own strategy and get a potentially higher payoff. As long as v is positive, the payoff v will be greater than $\frac{1}{2}v$. If one player plays Passive, the opponent is inevitably drawn to Aggressive, since v is the temptation payoff; Passive is left with the sucker payoff of zero, and he and his kind will die off. The reasoning here is parallel to the reasoning about Prisoner's Dilemma (see the discussion about figure 4.18).

But consider what will happen if v is greater than c. In other words, the value of the territory is greater than the cost of injury. In this case, the hawk-dove game really is equivalent to Prisoner's Dilemma; Passive is a dominated strategy, and a player shouldn't choose to play it.

This reasoning is reinforced if we consider a population of passive doves. Since playing Aggressive against Passive always wins, the dove population can easily be invaded by a single hawk. That hawk will have greater fitness than any dove it might play; so the hawks will reproduce at a greater rate, and eventually the doves will vanish. So the doves can be invaded by hawks.

Compare this with a population of aggressive hawks. Can doves invade this population? Clearly not. The doves will always fare worse than the hawks, and again they are doomed to extinction. It follows that hawks cannot be invaded by doves. It's clear that when $v > c$, the evolutionarily stable strategy, the ESS, is just (Aggressive, Aggressive). This is exactly what happens with a one-shot prisoner's dilemma game: only a sucker would cooperate in that case.

More formally, the expected utility of playing Aggressive is greater than the expected utility of playing Passive. Suppose that passive doves occur with probability d; aggressive hawks occur with probability $(1 - d)$, since there are only two types. The expected utility for playing Aggressive is

$$(d \times v) + \left[(1 - d) \times \frac{v - c}{2}\right].$$

The expected utility for Passive is

$$\left(d \times \frac{v}{2}\right) + [(1 - d) \times 0].$$

If $v > c > 0$, then $(v - c)/2 > 0$ and $v > v/2$. It follows that

$$(d \times v) + \left[(1 - d) \times \frac{v - c}{2}\right] > \left(d \times \frac{v}{2}\right) + [(1 - d) \times 0].$$

That is, the expected utility for playing Aggressive is greater than the expected utility for playing Passive.

Would Tit for Tat arise spontaneously in this case? No, because here the play is not iterated. When the same players play each other repeatedly, not knowing when they will play each other for the last time, both players do better cooperating. But in evolutionary game theory, a *population* plays once and then is reconstituted for the next generation. Each agent plays only once before fitness is computed and the next generation enters the scene. Chapter 9 discusses this topic in greater detail.

What happens when v, the value of the territory, is less than the cost of injury, c? Clearly, playing a hawk against another hawk is not very attractive. Since $v < c$, it follows that the payoff $(v - c)/2$ is negative. In this case, the hawk-dove game is the same as Chicken, where two drivers drive at each other, each one hoping that the other swerves first. No one wants to crash, but if a driver knows that the other will swerve first, he has every incentive to go straight ahead. This is essentially like the sidewalk game; a mixed strategy might work.

Now let $v < c$, so that $(v - c)/2 < 0$, and $v > 0$, which implies $v > v/2$. Suppose that a population is composed almost entirely of hawks, with just a small proportion of mutant doves, d. Since d is very small, any positive term multiplied by d will be smaller than if it is multiplied by $(1 - d)$. (In this population, one is more likely to encounter a hawk, after all.) Because of this, the expected utilities of dove and hawk are as follows (dove on the left side of the inequality, hawk on the right):

$$\left(d \times \frac{v}{2}\right) + [(1 - d) \times 0] > (d \times v) + \left[(1 - d) \times \frac{v - c}{2}\right].$$

The frequency of hawks in the population seems to reduce their fitness. The doves have higher fitness than the hawks and can therefore invade them successfully.

If the population is composed almost entirely of doves, with just a small proportion, h, of mutant hawks, then $(1 - h)$, the probability of being a dove, is appreciably larger than h, the probability of being a hawk:

$$[(1 - h) \times v] + \left(h \times \frac{v - c}{2}\right) > (1 - h) \times \frac{v}{2} + (h \times 0).$$

In this case, the hawks have greater fitness, so they can invade the doves. Thus, mutants of either type can invade a population of the other type. This means that neither pure phenotype is an ESS. The hawk-dove game shows that fitness can be *frequency-dependent*. The success of a strategy depends in some way on the frequencies of the other strategies in the population. The utility to an individual who plays Aggressive depends on the likelihood of encountering another individual who plays Aggressive.

What would the resulting population would look like? There are two possible outcomes when $v < c$. First, each player might play a pure strategy, but the population is a mix of the two types of players. Second, each player uses a mixed strategy.

Consider the case where there are two types of players, each playing a pure strategy. What is the optimal proportion of players? There are h hawks and $(1 - h)$ doves. Recall that the expected utility for a hawk is $h \times (v - c)/2 + (1 - h)v$, and the expected utility for being a dove is $[(1 - h) \times v/2] + (h \times 0)$. So it's better to be a hawk when

$$h \times \frac{v-c}{2} + (1-h)v > (1-h) \times \frac{v}{2},$$

or, simplified,

$$h < \frac{v}{c}.$$

The hawk type is fitter when the proportion of hawks is less than the utility of the territory divided by the cost of fighting.

It's better to be a dove if

$$h > \frac{v}{c}.$$

There is a balancing point

when $h = \frac{v}{c}$;

this is the mixed strategy Nash equilibrium.

What about the case where the agents play a mixed strategy? The mixed strategy equilibrium is $p = v/c$, the probability of playing a hawk. Let $1 - p$ be the probability of playing a dove; this probability comes out to $(v - c)/c$. Now we can briefly consider what happens when a mutant mixed strategy player invades populations playing a pure strategy.

Suppose a hawk meets a mixed strategy player. The latter acts like a hawk with probability p and acts like a dove with probability $(1 - p)$, so the expected utility for a hawk is

$$\left(p \times \frac{v-c}{2}\right) + (1-p)v.$$

By replacing p by v/c, this can be simplified to

$$v\frac{c-v}{2c}.$$

What about when a dove plays a mixed strategy player? In this case, the expected utility for a dove is

$$(p \times 0) + (1-p)\frac{v}{2}.$$

If p is replaced by v/c, the expected utility for a dove can be simplified to

$$v\frac{c-v}{2c},$$

the same as the expected utility for a hawk playing against a mixed strategy player. In fact, a mixed strategy player against a mixed strategy player gets exactly the same expected utility.

Although the argument is a bit thornier than the preceding one, it can be shown that neither a mutant hawk nor a mutant dove can invade a population of mixed strategy players. It follows that mixed strategy players have an ESS. Thus, when $v < c$, there are two stable outcomes: a mixed population of pure strategy players, or a population of mixed strategy players.

Let's return to the stag hunt game (see figure 4.22). The game was interesting because it had two distinct Nash equilibria—the payoff-dominant equilibrium of hunting deer as a group, and the risk-dominant equilibrium of hunting hare individually—and because Stag Hunt seems to model a linguistic phenomenon: drawing an implicature from an utterance, which corresponds to the payoff-dominant strategy.

The question arises whether a stag hunter can successfully invade a population of hare hunters. In other words, can a player playing a payoff-dominant strategy invade a population playing a risk-dominant strategy? Many people have the robust intuition that a stag hunter ought to be able to overthrow the hare hunters because the cooperative strategy pays better than defecting. But this is only true if the stag hunter can find other agents willing to play the payoff-dominant strategy. If a stag hunter plays a hare hunter, the stag hunter's payoff is 0, whereas the hare hunter's payoff is 1. Therefore, the hare hunter will appear to fare better than the stag hunter. If there aren't enough stag hunters in the population, their strategy will do less well than average, and they will disappear from the population.

In order for a payoff-dominant player to invade a population of risk-dominant players, there must be a sufficient number of payoff-dominant players to bring their payoffs above the population average. The question is, How did the payoff-dominant players get into a risk-dominant population? This raises interesting empirical questions about how cooperation (and reciprocity) could evolve. Intuitively, a population of cooperators should do better than a population of defectors, but how can enough cooperators get into the population to start the ball rolling?

Another example of the frequency dependence of fitness might be provided by grammatical attrition. As a boy, for example, I learned a grammar that allowed for constructions with a double modal, which expressed a subjunctive meaning:

(9) a. I *might could* do that.
   b. I *should ought* to have brought that book with me.

For some period of time, I was able to switch from one grammar to the other. That is, I used the double modal grammar with some probability p and the other grammar with probability $1 - p$. Of course, my utility for using the double modal grammar depended on whom I was talking to. I can model this by saying that I had acquired two grammars, one with double modals and the other a more standard grammar with single modals.

After I left west Texas, my expected utility for using the double modal grammar was sufficiently low that I eventually stopped using it. As it is with an individual, so it is with a population; if one grammar gets an edge in utility, it will tend to be used more. We can think of grammatical change as the competition between grammars for a resource; in this case, the resource is simply being used by an individual to express some message. A competing grammar will win, and supplant its competitor, if the expected utility of using that grammar is higher than the expected utility of the competitor, as with the hawk-dove game. Had I stayed in the southwest and not become an academic, the expected utility of my double modal grammar, relative to the standard grammar, might have been higher. Now, of course, I've mostly forgotten how my double modal grammar worked; my double modal grammar is a victim of grammatical attrition.

Case Marking Systems

Let's take an example of an application of evolutionary game theory to natural language typology. Jäger (2007) investigated the typology of case marking systems. This section reviews his work.

A noun phrase may be marked by a phonological element, a case marker, to show the grammatical role that the noun phrase plays in the sentence. English has very little in the way of morphological case, but it does mark pronouns:

(10) She was visiting her.

In the example in (10), the third person singular feminine pronoun occurs twice. First, it occurs as the subject of the sentence and emerges as *she*, its nominative form; second, it occurs as the object of the verb and emerges as *her*, its accusative form. In general, case marking systems are a way of disambiguating grammatical functions of noun phrases. English doesn't have much in the way of a case marking system, but it does use fairly strict word order to mark the grammatical functions of noun phrases. Languages with a lot of case marking tend to have freer word order, since the case markers show the function that the noun phrase plays in the sentence.

You can imagine all sorts of case marking systems that could potentially be used by natural languages. Simply take the cross-product of grammatical role, person, number, gender, animacy, definiteness, and specifity, and you could get a respectably large number of options. However, in the distribution of actual languages, only a few of these systems are frequent; many of the possibilities are extremely rare or unattested. In fact, most of the world's languages fall into either an *accusative system* with differential marking of objects, an *ergative system* with differential marking of subjects, or a mixed system that combines aspects of these two paradigms. A pure accusative system would add one or more morphological markers to objects, depending on the properties of the object noun phrase. In an ergative system, the subject of an intransitive gets special marking that distinguishes it from a transitive subject.

Why should these systems predominate? Jäger (2007) used evolutionary game theory to see whether there might be an economic explanation for this state of affairs. In broad terms, the idea is that the attested systems are viable solutions to reconciling two principles:

(11) **Speaker Economy**
Speakers strive to minimize effort in producing utterances.

Hearer Economy
Hearers prefer to minimize ambiguity.

Both principles involve reducing effort, either on the part of the speaker or on the part of the hearer. Notice that the two principles can come into conflict. The speaker might prefer to say less, leaving the hearer to puzzle out an ambiguous statement; the hearer might prefer the speaker to say more, thus making it easier for her to work out his intended meaning. This point is discussed more fully in chapter 6.

Speakers must make the effort to produce the appropriate case marker; without it, the hearer is faced with an ambiguity about the role that the noun phrase plays in the sentence. By the economy principles in (11), we would expect speakers to prefer to omit case markers and hearers to prefer that case markers be present.

The basic case marking systems are shown in table 4.1. The first column shows the case marking rules in schematic form, and the second

Table 4.1
Case Marking Systems

Case Marking Rules	A/p A/n O/p O/n
A → e, O → a	eeaa
A/p → e, A/n → z, O → a	ezaa
A/p → z, A/n → e, O → a	zeaa
A → z, O → a	zzaa
A → e, O/p → a, O/n → z	eeaz
A → e, O/p → z, O/n → a	eeza
A → e, O → z	eezz

Note: e = ergative; a = accusative; z = zero.

column shows the resulting markings for prominent subjects (A/p), nonprominent subjects (A/n), prominent objects (O/p), and nonprominent objects (O/n). When there is no prominence distinction, subjects are noted as A and objects are noted as O. By *prominence* is meant relative salience in a discourse; thus, pronouns are prominent, but indefinite noun phrases like *a man from Chicago* are taken to be nonprominent. Prominence is discussed in more detail in chapters 7–9.

In table 4.1, for example, the second row has three rules: prominent subjects are marked ergative, nonprominent subjects have no marking, and objects are uniformly marked accusative. The resulting system is abbreviated as *ezaa*. Notice that this abbreviation does not correspond to a word order frame; it simply encodes how case marking plays out for various types of subjects and objects. Ambiguity in the system can be avoided if at least one type of noun phrase is marked, even if the others are marked zero (no case marking). The strategies in table 4.1 are, in fact, the only systems that guarantee disambiguation between subjects and objects.

The idea here is that speakers and hearers are playing a game. Speakers encode a message using some case marking system, and hearers try to decode the message using a case marking system as a guide. If m is a message, then the speaker, s, maps from a message to a signal, and a hearer, h, maps from the signal back to a message. If all is working smoothly,

$$h(s(m)) = m.$$

To incorporate this into the definition of utility, it is quantified with the following function:

$$\delta_m(s, h) = \begin{cases} 1 & \text{iff } h(s(m)) = m, \\ 0 & \text{otherwise.} \end{cases}$$

So we can imagine a population of speakers, each using some case marking system specified in table 4.1 to encode her message, and a population of hearers, each using some case marking system to decode the message he receives.

Now, we need to encode the preferences of the players. Speakers have two possibly conflicting interests: they want to be understood, and they want to minimize their effort. We can capture this by the following utility function:

(12) **Speaker Utility**
$$U_s(m, s, h) = \delta_m(s, h) - (k \times \text{cost}(s(m))),$$

where m is a message, s is the speaker, and h is the hearer. That is, the payoff to the speaker is calculated by taking into account the success of the transmission of the message, $\delta_m(s, h)$, minus the cost of encoding the message times k, a constant that formalizes how important communicative success is for the speaker. Recall that explicitly marking a noun phrase incurs some cost, so the cost function simply returns the number of case-marked noun phrases in a clause.

If k is small, the speaker will tolerate a high cost for encoding the message. If k is large, the cost of encoding the message will rise. This means that assigning various values to k will result in a variety of different utilities.

What about the utility for the hearer? In this case, the hearer is unconcerned about the speaker's encoding costs and just wants to reconstruct the message that the speaker intended. The hearer's utility, then, is just the δ_m function:

(13) **Hearer Utility**
$$U_h(m, s, h) = \delta_m(s, h).$$

However, the real interest lies in calculating the expected utility of the various outcomes. The case marking systems in table 4.1 have different costs depending on properties of the message to be encoded. A prominent subject and a prominent object might require more marking than a prominent subject and a nonprominent object, for example. The expected utility should thus be a function of the likelihood of a message:

(14) **Speaker Expected Utility**
$$EU_s(s, h) = \sum_m P(m) \times (\delta_m(s, h) - (k \times \text{cost}(s(m)))).$$

Hearer Expected Utility
$$EU_h(s, h) = \sum_m P(m) \times \delta_m(s, h).$$

$P(m)$ can be approximated by estimating the probability of different classes of messages, that is, by counting the number of occurrences of different types of sentences in a corpus of actual text. We would need to count the number of occurrences of sentences with a prominent subject and a nonprominent object. By working out all the combinations of types, we could obtain a good approximation of the probability of different clause types. Jäger (2007) provided counts from Geoffrey Sampson's CHRISTINE corpus of spoken English as well as a hand-annotated subset of a corpus of spoken Swedish, the subset of the "Samtal i Göteborg" ("Conversations in Göteborg"), annotated by Östen Dahl.

Some of the strategies based on the systems in table 4.1 can be culled on the basis of utility. For example, the system *eeaa* is always strictly dominated because it uses two case markers for every transitive clause; one case marker would be adequate. To see this, compare *eeaa* with *eezz*. The latter is also unambiguous—the case-marked element is always the subject—and it costs only one unit per clause. Therefore, *eeaa* can be excluded from the analysis. This kind of analysis for all the systems in table 4.1 shows that *eeaa*, *eeza*, *eeaz*, *ezaa*, and *zeaa*, are all strictly dominated and can be excluded, since they needlessly mark both noun phrases in at least one clause type.

Before turning to the full analysis, let's consider the hearer's strategies. The hearer's strategy is using a method of mapping forms to meanings, where the only clues the hearer gets are the prominence values of the noun phrases and the word order. In addition, Jäger assumed a *faithfulness* constraint that requires that ergative mark A and accusative mark O. There are four possible clause types, depending on the order of the noun phrases:

- AO. The default order is Agent followed by Object.
- pA. If the order is in doubt, the most prominent element is the Agent.
- pO. If the order is in doubt, the most prominent element is the Object.
- OA. The default order is Object followed by Agent.

Table 4.2 shows an example game from Jäger in strategic normal form. The constant k that tunes the speaker's tolerance for disambiguation is quite low. The hearer's utilities can be derived by adding back the costs to each cell; since the hearer can only choose between cells in the same row, however, and the costs are identical within a row, the hearer's utilities can be identified with the speaker's utilities, so only the speaker's utilities are shown.

There are seven Nash equilibria in table 4.2, cells with the highest values in both their row and their column:

Table 4.2
Expected Utility Given Frequencies for Pronoun/Full NP in CHRISTINE ($k = 0.1$)

	AO	pA	pO	OA
eezz	0.90	0.90	0.90	0.90
zzaa	0.90	0.90	0.90	0.90
ezaz	0.85	0.85	0.85	0.85
zeza	0.81	0.81	0.81	0.81
zeaz	0.61	0.97	0.26	0.61
ezzz	0.86	0.86	0.87	0.86
zezz	0.54	0.89	0.54	0.54
zzaz	0.59	0.94	0.59	0.59
zzza	0.81	0.81	0.82	0.81
zzzz	0.50	0.85	0.15	0.50

zeaz/pA,
eezz/AO, eezz/pO, eezz/OA,
zzaa/AO, zzaa/pO, zzaa/OA.

Notice that the strategy pair zeaz/pA is the Pareto-dominant Nash equilibrium. This is quite reasonable and corresponds to a very common case marking system called a *split ergative system*; these systems combine ergative case marking with some aspects of accusative case marking. The other Nash equilibria are either pure ergative or pure accusative systems.

We can generate other games by changing the value for k—thus making speakers more or less interested in the communicative success of the utterance—or by changing the likelihoods of the various sentence types. Aside from k, however, we can estimate the expected utilities for the speakers and hearers using empirically measurable quantities.

We can now turn to the actual simulation. Since the case game has distinct strategies for speakers and hearers, assume that the population consists of two distinct subpopulations, one made up entirely of speakers and one made up entirely of hearers. The game is played between speakers and hearers, and the next round is constructed using *imitation dynamics*; players are occasionally offered to replace their strategy with the strategy of another player x. The probability that the player will adopt this new strategy is correlated with the success of that strategy in terms of utility. Clearly, more successful strategies will come to predominate in the population.

The game was played using the utility functions shown in table 4.2. The Pareto-dominant strategy zeaz/pA is an ESS. If there is no mutation, zzaa/AO and zzaa/pA are also stable; they cannot be invaded even by zeaz/pA mutants. Suppose *zeaz* players try to invade a zzaa/AO population; they will do much worse against *zzaa* speakers, since the latter will get a payoff of 0.90, while *zeaz* speakers get only 0.61. Thus, zeaz/pA mutants can't invade such a population. Other populations can be invaded, so in this case zeaz/pA, zzaa/AO, and zzaa/pA are the ESSs. These strategy pairs are attractors that pull the population toward them; the state that the system ends in will depend on the initial makeup of the populations, as was the case with hawks and doves.

The situation changes if mutation occurs. Suppose that occasionally an agent simply adopts a different strategy instead of copying a successful strategy. In this case, there is only one attractor in the system, the split-ergative system zeaz/pA. This means that given sufficient time, the only case marking system in the population will be zeaz/pA. That is, no how the system starts, whether at a pure strategy or a mixed strategy—a probability distribution over the pure strategies—the system will eventually be drawn to the attractor state, zeaz/pA. Because of mutation, the case-marking system zeaz/pA will eventually enter the population and, after that, the population will be drawn toward this system with the same inevitability that water in a basin is drawn to an open drain. This is an interesting result because split-ergative systems are quite common in the world's languages. There are, of course, quite a few other case marking systems; why aren't all languages drawn to the attractor state?

Notice that the players in the population are aware of the utilities of the game; this is what allows imitation dynamics to work. If this knowledge is hidden from the players, the system more closely resembles standard evolutionary game theory, where the agents are generally taken as unaware of the utilities, which are captured solely in terms of reproductive fitness. Jäger reported some simulations using this approach; in this case, with mutation, the stable strategies for speakers are *zeaz* (the split-ergative system), *zzaz* (differential object marking), *zezz* (differential subject marking), and *zzzz* (zero marking). Again, this is an interesting result because these systems exist and, in fact, represent the majority of languages. Jäger reported that these strategies are *stochastically stable*, which means that the equilibrium can withstand persistent random shocks, not just isolated shocks like an invasion by mutant forms. Stochastic stability is a stronger requirement that evolutionary stability (see Young 1998).

Let's return briefly to the question of how cooperation can arise in a population. How altruism and reciprocity, which are clearly related to cooperative behavior, came to exist is one of the great questions. A partial answer was provided by the biologist William D. Hamilton. His answer is related to a quip by J.B.S. Haldane, who, when asked whether he would give his life to save a drowning brother, replied, "No, but I would to save two brothers or eight cousins." Since a person shares one-half of his alleles with a brother and one-eighth of his alleles with a cousin, Haldane's answer gives the break-even benefit of altruistic behavior. Hamilton shared the intuition that altruistic behavior is more likely when the beneficiary is related to the altruistic agent.

Suppose that b is the benefit of altruistic behavior, and c is its cost. The payoff for an altruist act would be $b - c$. Hamilton reasoned that even if the altruistic act were quite costly to an individual, she might still provide some benefit to her genes if the recipient of the altruistic act were related to her. Let r be a measure of genetic relatedness; then we get the following precondition for altruistic acts, which is *Hamilton's rule*:

$rb > c$

That is, the amount of benefit accorded to your (potentially) related genes should exceed the cost. Of course, the notion of relatedness in Hamilton's rule is very hard to work out. Nevertheless, Hamilton's rule has inspired a great deal of work in biology.

Instead of directly working out r, suppose that the population is divided into cooperators and defectors. These labels are meant to reflect the prisoner's dilemma game, but they can also stand for the cooperators and defectors in a stag hunt game. Suppose we randomly pair up individuals and let them play a game, some variant of Prisoner's Dilemma, with b being the benefit of cooperating and c being the sucker payoff, which we can take to be the potential cost of cooperating.

To measure relatedness, let $\Pr(C|C)$ be the probability of a cooperator being paired with a cooperator, and $\Pr(C|D)$ be the probability of a cooperator being paired with a defector. Clearly, if I am a cooperator, I will meet either another cooperator or a defector. If I uniformly cooperate, then the likelihood of another cooperator's getting benefit is $\Pr(C|C) \times b$, and the likelihood of a defector's getting benefit is $\text{PR}(C|D) \times b$. So the net benefit to cooperators would be the benefit of aiding another cooperator less the benefit accorded to a defector: $(\Pr(C|C) \times b) - (\text{PR}(C|D) \times b)$. This simplifies to

$[\Pr(C|C) - \Pr(C|D)] \times b$.

By Hamilton's rule, cooperators should overtake defectors in the population when that quantity is greater than the cost c of cooperating, or

$$[\Pr(C|C) - \Pr(C|D)] \times b > c.$$

That is, it pays to cooperate when the number of cooperators is sufficiently large. This is the beginning of an account of how community can evolve; we are more likely to behave well in a world where others behave well.

There is a connection between linguistic behavior—the cooperation that occurs in signaling meaning and interpreting implicature—and the way such behavior can come about in a population. It seems to me that linguistic meaning shares many properties with ecology. Forms compete to occupy meaning niches just as species compete for ecological niches. In both cases, there is a combat for resources, albeit different kinds of resources. The mathematics needed for understanding these processes is the same in each case: a combination of game theory with evolutionary dynamics.

We have come, then, from the darkness of mutually assured destruction to the light of reciprocity and community. From the bleak world of Prisoner's Dilemma we see the first gleams of the evolution of goodness. Game theory, far from being a black art, provides a tool kit for examining how strategic interaction works. The tools themselves are only as good as those who wield them.

The following chapters discuss applications of game theory to various linguistic puzzles. A new horizon is visible: with game theory, we can think about the strategic aspects of language precisely and in a way that brings the study of language back to the broader forum of the social and behavioral sciences. Linguistic meaning and signaling have much in common with economics, anthropology, and mathematical ecology. The social aspects of strategic meaning can be understood, and doing so helps us to understand how language works and how we fit into the world.

I see in all the varieties of games something of the frenetic activity of the natural world. Language is not simply a static code but a complex network of forms—words, grammatical constructions—ever engaged in the interplay of meaning and competing with each other for use. With game theory, more than merely cataloguing these forms, we can study their lives, their use to express meaning and get things done. I can see in the games I build something of the life of language. Grammars provide a kind of bestiary, but the actual use—what we study by building games—shows the real social ecology of the language (see chapter 9). Linguistic forms are like organisms, competing with some, reinforcing others. Game theory is one of the keys to understanding the life of language.

I can well understand the sentiments expressed by the biologist W. D. Hamilton about his boyhood fascination with collecting and studying insects, moths, and butterflies. Late in his career he wrote movingly of this fascination:

What remains from all that fanatical activity? An odd and socially underdeveloped personality is probably the most conspicuous consequence as far as my friends and family are concerned. On the positive side there undoubtedly has also been the gradual induction of a rather vague and illogical brain into the endless fascination of science. But above both good and bad legacies from my own point of view it has helped me to carry on from child to adult a deep and never-ending gasp of wonder. First induced in me by the shining violet of a carabid's cuticle beneath my earliest stones, there lighting dark earth where woodlice crept and showers of Collembola leapt under my breath, then induced again and intensified by the deep velvet black and brilliant red and the white of a red admiral's wing close to my face on the flowers of my mother's Michaelmas Daisies, a long, long indrawn breath at the beauty of insects has stayed with me. I can still almost hear the hiss of the movement of those great wings, more beautiful in pattern on the underside than a Persian carpet, as they are raised and lowered and the butterfly basks in the sunshine on my mother's purple flowers. How clearly in imaginative memory I can still watch a Silver-washed Fritillary glide to a bramble flower through the heavy larch-scented air of our summer woods! (Hamilton 2001, 115)

Hamilton was one of the key figures in the early development of evolutionary game theory. He introduced many of the notions in a pair of papers published in 1964, in which he defined the notion of *inclusive fitness*. His 1967 paper on extraordinary sex ratios introduced the idea of an unbeatable strategy, which eventually developed into the ESS. He was also an early proponent of the Red Queen hypothesis that sex developed as a way of presenting parasites with constantly changing combinations of genes; parasites would respond by evolving new mechanisms to get around these defenses, and so on, in an endless arms race.

Hamilton died of a cerebral hemorrhage shortly after returning from a trip to the Democratic Republic of Congo, where he was conducting research. A few years before his death, he wrote,

Shivering a little I think of how, by the time I am old, all these secrets of their work will be known, of how easily, then, we will super-attract beetles if we care to from large areas of forest by means of foetid chemicals... I think how, by that time, I can confidently arrange what I have thought of. I will leave a sum in my last will for my body to be carried to Brazil and to these forests. It will be laid out in a manner secure against the possums and the vultures just as we make our chickens secure; and this great Coprophanaeus beetle will bury me. They will enter, will bury, will live on my flesh; and in the shape of their children and mine, I will escape death. No worm for me nor sordid fly, I will buzz in the dusk like a huge bumble bee. I will be many, buzz even as a swarm of motorbikes, be borne, body by flying body out into the Brazilian wilderness beneath the stars, lofted under those beautiful and un-fused elytra which we will all hold over our backs. So finally I too will shine like a violet ground beetle under a stone. (Hamilton 2001, 122)

What a strange and lovely image of immortality.

Further Reading

Game theory has its roots in the work of the French mathematician Émile Borel, who wrote a paper in 1921 on games like poker that involve elements of chance along with bluffing behavior. Borel was able to lay out many elements of the mathematics but did not manage to prove many fundamental results. His work was extended by John von Neumann and later by von Neumann and Oskar Morgenstern in their book *Theory of Games and Economic Behavior* (1944), which established game theory as an independent field. Their work was largely concerned with zero-sum games, purely competitive games, with a clear winner and a clear loser, in which the players' payoffs sum to zero.

People trained in the humanities often find the mathematics of game theory daunting, but many good introductions to game theory are available, and the basic ideas can be grasped even when stripped of equations. Certainly, no one should rest content with the brief treatment that I provide. A good place to start is Binmore (2007), which has the advantage of brevity and thrift. Dixit and Nalebuff (1991; 2008) and Dixit and Skeath (2004) provide lively and accessible introductions to game theory.

I have relied throughout on Osborne's (2004) excellent introductory textbook and have also consulted Myerson (1991) and Osborne and Rubinstein (1994). Luce and Raiffa (1957) provide an indispensable discussion of utility in game theory. Benz, Jäger, and van Rooy (2006) give a general introduction to game theory intended for linguists. Poundstone (1992) has been another source, particularly for historical material. I also consulted Macrae's (1992) biography of von Neumann. Poundstone provides the source for the cake game, although my discussion is rather different from his. The Holmes-Moriarty game is from von Neumann and Morgenstern (1944). Glimcher (2003) discusses a game-theoretic treatment of dining ducks. Schelling's work is extremely approachable. Schelling (1960) is one of the great classics of game theory and possibly the most influential book in the social sciences. Schelling (2006) is a representative collection of essays. Schelling (1978) has been a great source of inspiration, particularly for the work on convention formation and meaning as an emergent property of social systems, which I discuss in chapter 9.

The hobo dinner game is my version of the battle of the sexes game, familiar from introductory game theory; Osborne (2004) and Osborne and Rubinstein (1994) call it Bach or Stravinsky? I apologize to all hobos if my presentation is insensitive. The discussion of the connections between mixed strategies and language variation is largely mine; I owe a great deal to Bill Labov, in particular Labov (1994), for enriching my thinking about variation. My thinking about coordination games and implicature owes something to Sally (2003).

The prisoner's dilemma game can be found in virtually every introductory book on game theory. Axelrod (1984) and Axelrod and Hamilton (1981) connect the game to evolutionary theory in a very direct way. The various versions of the sidewalk game as instances of prisoner's dilemma are mine and reflect my puzzling about pedestrians in Philadelphia and New York.

My discussion of the stag hunt game owes a great deal to Skyrms (2004). The discussion of risk dominance and payoff dominance owes much to the work of Harsanyi and Selten (1988), although I was led to it by Sally (2003). The stag hunt arms race is from Osborne (2004). Parikh (2001), in his application of game theory to linguistic pragmatics, argues for Pareto dominance (payoff dominance).

The core text in evolutionary game theory is probably Hofbauer and Sigmund (1998). Sigmund (1993) is a very readable discussion of some of the core ideas, and Nowak (2006) gives a more technical but still approachable introduction to the field. For the mathematically inclined, Gintis (2000) is a good introduction to game theory from the point of view of evolution. Evolutionary game theory owes much to the work of Hamilton (1964; 1967) and Maynard Smith (1972; 1982). My discussion of the application of evolutionary game theory to social evolution is indebted to both Skyrms (2004) and McElreath and Boyd (2007). Some basic works on the evolution of convention in economics include Kandori, Mailath, and Rob (1993) and Young (1993). Grammar competition has been developed and discussed by my colleague Tony Kroch; see, for example, Kroch (1989).

5 A Game Logic for Natural Language

Chapter 1 put forward the notion that the meaning of a sentence involves truth. We know what a sentence means because we know what the world would be like if the sentence were true. This line of thinking suggests that we need a method of working out the truth conditions of sentences, descriptions of what the world must be like if the sentence is true. The truth conditions of a sentence can be computed using a zero-sum game of perfect information, essentially the same kind of game as the cake game (see chapter 4). In fact, we can use backward induction on the game to decide whether a given sentence is true or false.

This chapter presents a first-order logic whose formulas look like sentences of English. By following the examples, readers should become fairly familiar with first-order logic. The examples show how logic can be reconstructed as a game. Certainly, one can describe all sorts of odd logical constructs as games, but working out how to describe such niceties would take us too far afield. Let's stick to the comforts of the Aristotelian world.

There is certainly something to the idea behind truth conditions. I use language to transmit information to others about the world, at least, what I *think* the world is like. It makes sense to say that this information is at least partly characterized by a description of what the sentence is claiming about the world. On the other hand, truth conditions are an awfully thin hook on which to hang meaning. Look at the following pair of sentences:

(1) a. Barack Obama is the president of the United States.
    b. The president of the United States is Barack Obama.

It's hard to imagine how the first sentence, (1a), could be true and the second sentence, (1b), could simultaneously be false. Both sentences say the same thing about the world. Nevertheless, the two sentences *do* different things.

Suppose someone asked the question, "What's Barack Obama up to these days?" Answering the question with (1b) would be extremely peculiar because we're talking, after all, about Obama. A more natural response would be,

(2) He's the president of the United States.

which is more or less parallel to (1a). If, on the other hand, we were discussing the United States—its political system, for instance—we might quite naturally answer,

(3) The (current) president of the United States is Barack Obama.

Parentheses around a word, like the parentheses around *current* in (3), mean that the word or phrase inside them can be omitted or included. In the case of (3), we're talking about the United States and indicating something about its current president. In other words, the two sentences (1a) and (1b) might be true in exactly the same circumstances, but they're *doing* completely different things.

What about the truth conditions for a suggestion? When is the following sentence true?

(4) You might try turning the knob counterclockwise.

It seems more accurate to say that a sentence like (4) is doing something. Comparing (4) with a command like

(5) Turn the knob counterclockwise!

suggests that (4) is a polite form of the command in (5). One can argue, of course, that understanding utterances like (4) and (5) will involve some grasp of truth conditions—I need to know what it means to turn some particular knob counterclockwise—but that's different from supposing that the meaning of a sentence is given entirely by its truth conditions.

All this suggests that while the meaning of a sentence may involve truth conditions (or appropriateness conditions; see chapter 9), truth conditions do not exhaust meaning. More broadly, we can distinguish between a truth-conditional semantic component—which can be worked out via equilibrium strategies of zero-sum games of perfect information—and a non-truth-conditional component, which can be worked out as a coordination game between a speaker and a hearer.

This chapter explains the process of defining the truth conditions of sentences; bear in mind that truth conditions are only one component of meaning. The method given here for working out truth conditions uses extensive zero-sum games of perfect information with a clear winner and a clear loser.

Most current approaches to truth-conditional semantics use the formalisms and methods of logic to provide an account of semantic meaning. On the one hand, this seems to be a sensible move, since logic and model theory provide both formal rigor and a certain mathematical elegance to the analysis of meaning. On the other hand, these accounts use the methods of artificial languages, like logic, to give an account of meaning; indeed, Richard Montague, possibly the most important early innovator in the formal semantics movement, titled one of his seminal papers "English as a Formal Language." Tarski's work on logic, on which Montague relied heavily, was intended to give an account of truth in a formal language.

I would argue that it is worthwhile to think in the opposite direction; instead of moving from logic to natural language meaning, one should think about how logic could arise from the ordinary use of natural language. Most ordinary language use is cooperative; logic is what one gets when language becomes adversarial.

I follow the broad outlines of the method developed by Tarski and Montague. As many theorists of compositional semantics do, I start by giving a definition of the syntax of the language; this is just a grammar. I outline a particular formalism, *tree-adjoining grammar* (TAG), that is intuitive and easy to develop. I then describe a structure called a model, a mathematical structure that can be taken as a simulacrum of the actual world. Any sentence can then be evaluated with respect to the model.

This is usually done by writing an abstract function that maps things in the language to things in the model. The truth of a sentence with respect to a model can then be computed using this function. I do things slightly differently here. Instead of giving a function that interprets linguistic symbols relative to a model, I give a set of game rules that define games that can be played on these symbols. This approach to semantics, called *game-theoretic semantics*, was initially developed by the philosopher Jaakko Hintikka and his colleagues, although it has antecedents far back in the history of logic. In earlier forms of logic, logical reasoning was thought of as a contest between an advocate, who sought to support, and a preceptor, who offered troublesome counterexamples.

Imagine two players, Abélard and Eloïse, who are playing a game of verification and falsification. For present purposes, these players are abstract agents; this is still a far cry from the social system of meaning that I argued for in part I, but it will help to show how games can be used to model linguistic processes. More realistic social games are discussed in part III.

Nevertheless, I think there is great value in working with game-theoretic semantics. These games all involve picking and choosing over a model. They are small laboratories for thinking about how the meanings of linguistic expressions can be constructed by strategic interaction with the world. Indeed, one might think of formal logic as developing out of conversational practice by the development of adversarial games out of ordinary cooperative games. The work in this chapter and the next is crucial for understanding the broad question of the social and strategic aspects of meaning.

Both players are rational; they are endowed with knowledge of the model, and they know and understand the game rules. The two players are, in fact, playing a zero-sum game, a game of absolute competition with a clear winner and a clear loser. Suppose Eloïse has bet that the model supports the sentence; this means she has wagered that the sentence correctly characterizes the model. Abélard, on the other hand, has wagered that the model does not support the sentence. Eloïse will try to find support for the sentence, while Abélard will try to find counterexamples.

The Tale of Abélard and Eloïse

Since I use the characters Abélard and Eloïse in the development of game logic, it's worth taking a moment to describe the historical figures on which they are based. Peter Abélard (French: Pierre Abélard) (1079–1142) and Eloïse (ca. 1098–1164) are probably best known today for their correspondence, which is both passionate and philosophical. Abélard was in his day a successful academic who was a major proponent of nominalism, the doctrine that universals don't exist. This is one reason I base a character on him in the games; the character's role is trying to find counterexamples to universal statements. Eloïse was a young woman, the ward of her uncle Fulbert, who was not only beautiful but an accomplished classicist. Abélard immediately set about worming his way into her good graces.

Soon enough, he became her tutor and set about seducing her. Eventually, Eloïse's uncle Fulbert began to suspect this and, what's worse, Eloïse became pregnant. Abélard managed to spirit her off to stay with his sister, and she gave birth to a son, whom she named Astrolabe after one of the technology marvels of the age. I find this detail quite endearing, although it is a bit like naming one's child Laptop.

Eventually, Eloïse and Abélard were married, whereupon Abélard sent her off to a nunnery. Needless to say, Uncle Fulbert was not happy; he concocted a terrible vengeance:

> Violently incensed, they laid a plot against me, and one night, while I, all unsuspecting, was asleep in a secret room in my lodgings, they broke in with the help of one of my servants, whom they had bribed. There they had vengeance on me with a most cruel and most shameful punishment, such as astounded the whole world, for they cut off those parts of my body with which I had done that which was the cause of their sorrow. This done, straightway they fled, but two of them were captured, and suffered the loss of their eyes and their genital organs. One of these two was the aforesaid servant, who, even while he was still in my service, had been led by his avarice to betray me. (Abélard 1922, 29–30)

If Uncle Fulbert thought he could save Eloïse from the convent, he was quite mistaken. She took the vows and carried on a passionate epistolary romance with Abélard.

Syntax

The meanings of artificial languages, like the language of first-order logic, are defined by specifying first the syntactic structure of the language and then the interpretation of the resulting signs relative to a model-theoretic structure. In essence, the grammar of the language is given, and a set of rules maps the resulting forms onto the world represented by the model.

In this chapter, I create an artificial language that looks very much like a fragment of English. In fact, this language is an artificial one that permits the expression of meanings that are compatible with first-order logic. I define some semantic rules that when interpreted as extensive games of complete information (like the cake game) allow the semantic interpretation of the language. The resulting system will give readers a game perspective on truth-conditional semantics.

Figure 2.1 showed an example of a context-free phrase structure grammar. I would be surprised to find any linguist who currently believes that a context-free phrase structure grammar is powerful enough to fully describe the syntactic patterns found in natural language, so I explore here a kind of grammar that might be powerful enough to do so.

Recall, in chapter 2, the method of constructing parse trees using rewrite rules. A set of symbols was specified, and the rules told how to rewrite those symbols to generate sentences. The framework in this chapter,

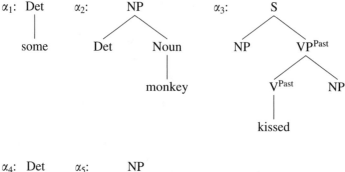
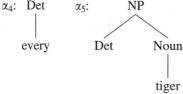

Figure 5.1
A Small Tree-Adjoining Grammar

tree-adjoining grammar, takes parse trees to be primitive elements of the system. A set of trees is specified that forms the basis of the system.

Consider the following sentence:

(6) Some monkey kissed every tiger.

Each word in the sentence has an elementary tree structure associated with it (figure 5.1).

The first tree in the figure, labeled α_1, is a very simple tree whose root node is Det and whose leaf node is *some*. The second tree, labeled α_2, has a node Det as one of its leaf nodes. The operation of substitution identifies the root node of α_1 with the matching leaf node in α_2. This operation yields

(7)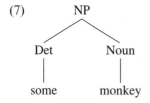

The result in (7) is a tree that is itself available for substitution. The tree in (7) can be substituted for the subject NP node in the elementary tree, α_3. The result is

(8)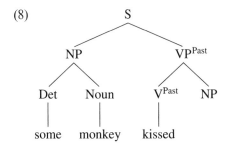

The tree α_4 can be substituted into α_5 by identifying its root node, Det, with the frontier Det node in α_5. This gives

(9)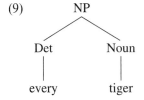

Now the tree in (9) can be substituted into the tree in (8) to get

(10)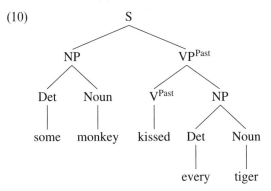

The yield of the tree in (10)—its leaf nodes read from left to right—is the sentence *some monkey kissed every tiger*.

Tree-adjoining grammars have another operation called adjunction. Suppose we want to derive the sentence

(11) Some monkey didn't kiss every tiger.

Assume we have the elementary trees shown in figure 5.1 and want to supplement those with the two elementary trees in figure 5.2.

The first tree in figure 5.2, labeled α_6, has as a leaf node the negative element *didn't*. I have not assigned a category to *didn't*; I treat it here as a special logical element for negating sentences and show later how to

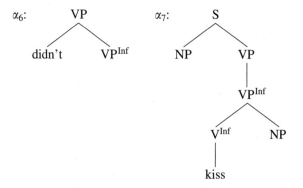

Figure 5.2
Elementary Trees

treat it as a semantic game. Notice that this tree is available to combine with a subtree marked VP^{Inf}. I mean by this notation a verb phrase headed by a verb that is not marked for tense. Such a subtree occurs as part of α_7 in figure 5.2.

Now let's turn to the adjunction operation itself. Suppose we break the tree α_7 into two pieces:

(12)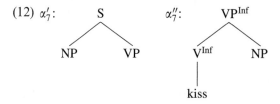

The tree α_6 can be substituted into α_7',

(13)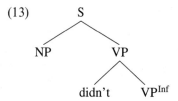

and the tree α_7'' can be substituted into the tree in (13) to get

A Game Logic for Natural Language

(14)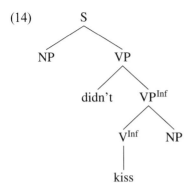

Now substitute the tree in (7) into the tree in (14):

(15)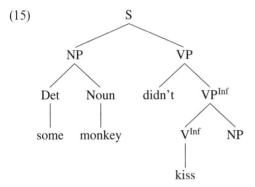

Finally, substitute the tree in (9) into the tree in (15) to get

(16)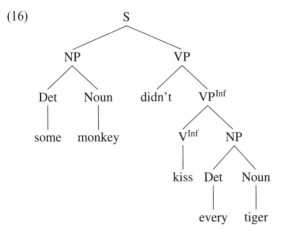

which yields *some monkey didn't kiss every tiger*.

The operation of adjunction, illustrated here, is sufficient to increase the power of the grammar beyond the context-free phrase structure system. The additional power seems to be necessary to account for the kind of complexity we observe in real language data. The adjunction operation, for example, is used to construct sentences involving long-distance dependencies like that found in some questions:

(17) Who did John think Mary visited?

The *wh*-word *who* in (17) is related to the object position of the verb *visited*. The example can be built by combining two trees. First, obtain the tree in (18):

(18)
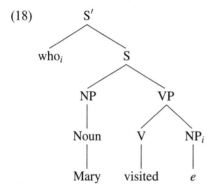

In order to show that the *wh*-word *who* is associated with the object of *visited*, I use the device of an index i on *who* and on an empty NP in object position. This last element shows the position with which *who* is semantically associated. Then adjoin the tree in (19) into the tree in (18):

(19)
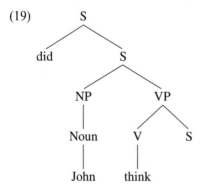

The result is shown in (20):

(20)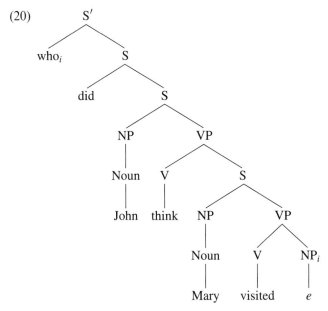

I don't give a deep analysis here, since I'm primarily interested in developing a logic, so I use the system to construct a formal language that looks rather like English.

The basic idea is to get a set of elementary trees that could be used to generate a language via the operations of substitution and adjunction. Each elementary tree will be associated with a game between Eloïse and Abélard. Some of the games might involve just looking at the model to verify whether something holds. Other games will involve strategically picking expressions to substitute for other expressions.

Figure 5.3 shows elementary trees for some logical operators. For the rest, assume that common nouns like *tiger* are generally associated with a tree like

where X can be filled in by any common noun. Names like *Mary* are associated with

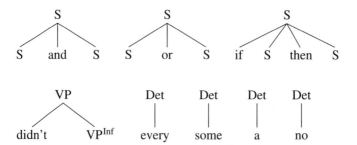

Figure 5.3
Elementary Trees for Some Logical Operators

where Y is filled in by a name.

Intransitive verbs like *walk* are associated with two elementary trees, one for the tensed case (*walked*) and one for the infinitive (walk):

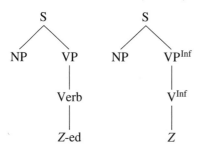

where Z is filled in by any intransitive verb. Analogously, transitive verbs like *chase* have trees like

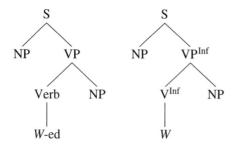

> **True and False (A Definition)**
>
> A sentence is true on a model if Eloïse has a winning strategy given that model, and a sentence is false on a model if Abélard has a winning strategy given that model.

Figure 5.4
True or False in a Model

where W is filled in by any transitive verb. Having two different structures for verbs makes it possible to build negated and positive sentences.

This should provide enough syntax to build some interesting sentences and to show how Abélard and Eloïse can play some games with them. The project now is to show how zero-sum games of perfect information can be used to explore truth conditions.

Games and Models

Let's now spell out the central idea of game-theoretic semantics. A set of game rules is specified that, when combined with a model, will specify a game tree for each grammatical sentence in a model fragment. In other words, Abélard and Eloïse will, when given a model and a grammatical sentence, be able to play an extensive game of perfect information based on that model. If Eloïse has a winning strategy on that sentence, then the sentence will be *true*; her winning strategy (or strategies) will stand in for the truth conditions of the sentence. On the other hand, if Abélard has a winning strategy on the sentence, then the sentence will be *false* with respect to that model. This important point is highlighted in figure 5.4.

The following notation indicates a game on a sentence, S, given a model, M:

$G(S; M)$.

Assume that S is actually given in the form of a parse tree. This isn't the usual assumption in game-theoretic semantics, but it makes some aspects of the presentation easier.

Atomic Sentences

We need a set of game rules that will allow Eloïse and Abélard to pick and choose among elements of the language. In a particular sentence generated by the grammar, for example,

(21) John saw a monkey and every tiger chased a boy.

Some elements are logically inert. For example, *John* is a name that picks out some individual. Assume that Eloïse and Abélard can simply look in the model and see the individual that is denoted by *John*. Equally, given a sentence like

(22) Alice saw Mary.

Eloïse and Abélard can simply look in the model and see whether the individual denoted by *Alice* actually did see the individual denoted by *Mary*. In other words, Eloïse and Abélard have perfect knowledge of the individuals in the model and the relations they stand in. The sentence in (22) is a special kind of sentence called an atomic sentence, defined as follows:

(23) **Atomic Sentence**
A sentence S is an atomic sentence if and only if it contains only names (proper nouns) and a single verb.

So the sentence in (22) is an atomic sentence according to the definition in (23), but the sentence in (21) is not because it contains elements other than proper nouns and a single verb.

The first game rule is as follows:

(24) **(R.atom)**
Suppose the game is $G(S; M)$. S is an atomic formula. The current verifier wins and the current falsifier loses if S holds in M. The current falsifier wins and the current verifier loses if S does not hold in M.

This rule says that to determine who wins on a particular sentence, the players need only look in the model. If the atomic sentence consists of a name and an intransitive verb, they simply look to see if the entity denoted by the name has the property named by the intransitive verb. If the atomic sentence consists of a pair of names and a transitive verb, the sentence holds if the pair of entities denoted by the names stands in the relation named by the transitive verb.

Eloïse and Abélard are the initial verifier and falsifier, respectively; they can change roles (cross-dressing?) under certain conditions. If the current verifier wins, and the current verifier is Abélard, that means the original sentence is false.

To show how the rule in (24) works, table 5.1 gives a small model. The linguistic expressions are in the first column, and their correspond-

Table 5.1
A Small Model

Expression	Model
John	j
Alice	a
Bill	b
slept	j, b
walked	a
snored	b
saw	$\langle a, j \rangle$

ing model-theoretic objects are in the second column. The entities in the model are represented by lowercase letters. An intransitive verb denotes a set of entities, and a transitive verb denotes a pair of entities that stand in the relation named by the verb; such a pair is represented by lowercase letters in angle brackets.

The world of table 5.1 has only three entities, named by *John*, *Alice*, and *Bill*. John and Bill slept while Alice walked, Bill snored, Alice saw John, and nothing else was happening. We can think of the model as a small part of the world relevant to playing some language game or games.

Now, let's see who wins on the following games:

(25) a. Alice snored.
   b. Bill slept.
   c. Alice saw John.
   d. John saw Alice.

Example (25a) asserts that Alice snored, but the model says that the only snorer is Bill. So (25a) doesn't hold, and the falsifier (namely, Abélard) wins. This means that (25a) is false in the model given in table 5.1.

Example (25b) asserts that Bill slept. The model in table 5.1 supports this, so the current verifier wins. Since the current verifier is Eloïse, the sentence is true.

Example (25c) asserts that Alice saw John. The model says that the entity named by *Alice* does indeed stand in the relation named by *saw* to the entity named by *John*. Thus, the model supports this, and the current verifier, Eloïse, wins; the sentence is true.

What about example (25d)? It asserts that John saw Alice, but in the model j does not stand in the *saw* relation to a; while Alice may have

Table 5.2
Negation

Sentence	Sen[didn't]tence
true	false
false	true

seen John, John didn't see Alice. So the model in table 5.1 does not support the sentence, and the current falsifier, Abélard, wins.

The idea behind the game rule in (24) is simply that the truth of an atomic sentence is a matter of looking at the world and immediately verifying whether the sentence holds or does not hold. Of course, simply looking to see whether something holds requires no strategic thinking on the part of the players; they're not choosing anything. Other elements of the language, like *every*, *and*, or *didn't* are logically active in the sense that Eloïse and Abélard must make strategic choices in order to determine who wins on a sentence containing those elements. In the sentence in (21), the following words and phrases are logically active: *a monkey*, *and*, *every tiger*, *a boy*. In each of these cases, a game rule must be specified for the logical agents to use in making strategic choices given the model.

Negation

The terminology in **(R.atom)** is a bit peculiar, since it refers to *the current verifier* instead of Eloïse, and *the current falsifier* instead of Abélard. The reason is that there might be cases where Eloïse and Abélard exchange roles. For instance, suppose Eloïse and Abélard are playing on the sentence

(26) Alice didn't snore.

Under what conditions would Eloïse win on (26)? Intuitively, she wins on (26) if the model fails to support

(27) Alice snored.

The truth value of a negated sentence and its corresponding positive sentence should flip. Table 5.2 indicates the correspondence between a sentence and its negation as "Sentence" and "Sen[didn't]tence."

Given a negative sentence like (26), Eloïse and Abélard should exchange roles and continue play on the corresponding positive sentence. This suggests the following game rule for *didn't*:

A Game Logic for Natural Language   141

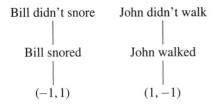

Figure 5.5
Two Games with Negation

(28) **(R.didn't)**
Suppose the game is $G(S; M)$, and S is

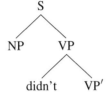

Then the current verifier and the current falsifier exchange roles, and the game continues as $G(S'; M)$, where S' is

The rule in (28) works by removing the negation and having the verifier and the falsifier change roles on the resulting positive sentence. I've fudged the rule in (28) a bit because the verb in the resulting positive sentence should be marked in the past tense for this fragment of English.

Figure 5.5 shows two extensive game trees, similar to the extensive game tree for the cake game (see figure 4.4). The sentences are

(29) a. Bill didn't snore.
     b. John didn't walk.

Inspection of the model in table 5.1 reveals that Bill snored and John didn't walk. Thus, example (29a) should be false (Abélard should win), and example (29b) should be true (Eloïse should win).

The game trees in figure 5.5 are exceedingly simple, since no real strategic choices are made. The information states are sentences. The root information state (at the top) is the original sentence, which gets replaced with the positive sentence.

To understand the payoffs, recall that Eloïse and Abélard are playing a zero-sum game; their payoffs sum to 0. Assume that the winner of a game gets a payoff of 1 and the loser accordingly gets a payoff of -1. Their payoffs are shown as a pair:

(Eloïse, Abélard).

Also recall that Eloïse and Abélard exchange roles in the move from the negated sentence to the positive sentence. When they exchange roles, Eloïse wants to falsify the sentence and Abélard wants to verify it. When the sentence in (29a) is replaced with *Bill snored* and the players exchange roles, Eloïse wants to falsify the sentence and Abélard wants to verify it. Since the sentence *Bill snored* holds in the model, Eloïse loses and Abélard wins. Since Abélard has a winning strategy, the sentence in (29a) is false.

Equally, when the sentence in (29b) is replaced by the positive *John walked*, Eloïse becomes the falsifier and Abélard becomes the verifier. Since *John walked* doesn't hold in the model in table 5.1, Eloïse wins and Abélard loses. Since Eloïse has a winning strategy on (29b), the sentence is true. The reader can check by simple inspection that the winning strategy can be worked out via backward induction on the game tree.

Logical Connectives

Let's look at some simple examples where Eloïse and Abélard will have to exercise strategic thinking. I discuss sentence connectives like *and* and *or*, and then the peculiar case of conditional sentences, keeping the interpretation close to standard logic both for simplicity and because that may just be the right semantic interpretation for them.

Conjunction In normal usage, *and* often gets a temporal interpretation; compare (30a) with (30b):

(30) a. John and April got married and they had a baby.
    b. John and April had a baby and they got married.

These examples show that it is normal to interpret the two sentences connected by *and* as temporally ordered. So (30a) is taken as saying something quite different from (30b). Equally, there might be a causal link between the two sentences, as in

(31) The president entered the room and the soldiers stood at attention.

where the soldiers stand at attention because the president is the commandin-chief of the armed forces.

Table 5.3
Logical *and*

Sentence$_1$	Sentence$_2$	Sentence$_1$ and Sentence$_2$
true	true	true
true	false	false
false	true	false
false	false	false

I consider *and* here in its logical use; its other interpretations arise from the context by rational decision making. I'll leave that problem for future work. The logical use is illustrated by (32); as (32a) and (32b) show, the order of the two sentences doesn't matter with logical *and*; both sentences are true:

(32) a. Mercury is a planet and the Sun is a star.
   b. The Sun is a star and Mercury is a planet.

As table 5.3 shows, a sentence made up of two sentences joined by *and* is true if both subsentences are true. In other words, in order for Eloïse to win on a sentence containing *and*, both subsentences must hold in the model. Abélard will win if just one of the subsentences fails to hold.

This suggests an easy game rule. Let the falsifier pick one of the subsentences to continue the game on. If the falsifier has a winning strategy on that subsentence, then he has a winning strategy on the whole sentence, which must be false. Here is the rule:

(33) **(R.and)**
Suppose the game is $G(S; M)$, and S is

Then the current falsifier picks a number i, where i is 1 or 2. The game then continues as

$G(S_i; M)$.

The rule **(R.and)** in (33) says that the falsifier gets to choose a subsentence to play on. If he can choose one where he has a winning strategy, then the whole sentence must be false.

The rule in (33) has the property that it takes a complex structure and replaces it with something simpler. All the rules in the system will have

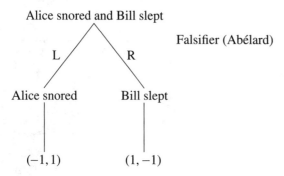

Figure 5.6
Extensive Game Tree for a Simple Conjunction Game

this property; logically active elements will be removed, and one player or the other will make a choice. Eventually, the complex sentence will be reduced to an atomic sentence, which can be verified or falsified by looking at the model.

Let's look at a concrete example using the model in table 5.1. Suppose that Eloïse and Abélard are playing on the sentence

(34) Alice snored and Bill slept.

The two game rules (**R.and**) in (33) and (**R.atom**) in (24) plus the model in table 5.1 are enough to determine an extensive game of perfect information.

Figure 5.6 shows the extensive game tree. The root of the tree shows the whole sentence, *Alice snored and Bill slept*. The only player with a move is Abélard, who may select either L or R; that is, he can select either the left conjunct, *Alice snored*, or the right conjunct, *Bill slept*. Since *Bill slept* holds in the model, but *Alice snored* doesn't, the utilities are as follows:

$u_{\text{Abélard}}(\text{Alice snored}) = 1.$

$u_{\text{Eloïse}}(\text{Alice snored}) = -1.$

$u_{\text{Abélard}}(\text{Bill slept}) = -1.$

$u_{\text{Eloïse}}(\text{Bill slept}) = 1.$

Clearly, Abélard's best choice, which is also an equilibrium, is to choose L. He has no reason to defect from this strategy; his strategy profile then is

{(Alice snored and Bill slept, L)}.

Table 5.4
Logical *Or*

Sentence₁	Sentence₂	Sentence₁ or Sentence₂
true	true	true
true	false	true
false	true	true
false	false	false

Given the structure of the game, Eloïse doesn't get to move. The equilibrium strategy gives the game to Abélard, so the sentence is false. Once again, rollback gives the correct result.

There are two things to notice about this example, simple though it is. First, the game emerges from the local model and the choices determined by the given semantic rules. Second, "true" and "false" are given by the Nash equilibrium of the game. If Eloïse wins in the Nash equilibrium strategy, then the sentence is true; otherwise, if the game goes to Abélard, the sentence is false. Thus, truth is analyzed in terms of the Nash equilibrium of a zero-sum game of perfect information.

The kind of game used here is quite simple, so one might want to make various changes. However, the notions presented here are sufficient to characterize first-order logic. Small changes will lead to different kinds of logic. In particular, if the game has cases where there is no clear winner, we get a multivalued logic. If the game is one of incomplete information, we get a new kind of logic, *independence-friendly logic*, which has been the subject of a great deal of research in recent years.

Disjunction Let's consider sentences joined by *or*. As before, *or* in everyday language can have a variety of interpretations. The following is uttered by a mother to her child:

(35) You can have ice cream or you can have cake.

In the case of example (35), she probably means that the child can have either ice cream or cake, but not both. This is very different from the case of a waiter uttering,

(36) Can I bring you drinks or an appetizer?

In this case, the waiter would be perfectly happy if I ordered both drinks and an appetizer.

The interpretation of *or* shown in table 5.4 is closer to what the waiter means. A sentence like *John walked or Bill slept* holds in a model if *John walked* holds or if *Bill slept* holds or if both sentences hold. The sentence is false only when both disjuncts fail to hold in the model. In order to verify a sentence with *or*, it is only necessary to verify one of the sentences it connects. Thus, there are more ways to verify such a sentence than to falsify it. This fact suggests the following game rule for *or*:

(37) **(R.or)**
Suppose the game is $G(S; M)$, and S is

Then the current verifier picks a number i, where i is 1 or 2. The game then continues as

$G(S_i; M)$.

In other words, to win on a disjunction, the verifier simply has to demonstrate that one of the disjuncts holds.

The rule can be illustrated with the following disjunction:

(38) Alice snored or John slept.

Once again, the model is the one in table 5.1. The truth of example (38) hinges on whether one of *Alice snored* or *John slept* holds in the model. Inspection reveals that the latter, but not the former, holds, so we have the following utilities:

$u_{\text{Abélard}}(\text{Alice snored}) = 1$.

$u_{\text{Eloïse}}(\text{Alice snored}) = -1$.

$u_{\text{Abélard}}(\text{John slept}) = -1$.

$u_{\text{Eloïse}}(\text{John slept}) = 1$.

The extensive game tree is shown in figure 5.7. The information state at the root of the game tree shows that Eloïse and Abélard are playing a game on the sentence *Alice snored or John slept*. By the rules of the game, the only move available is Eloïse's choice of the left or right disjunct. If Eloïse chooses the left disjunct, the game ends immediately and Abélard wins. If she chooesees the right disjunct, the game also ends immediately but she wins. So Eloïse has a winning strategy on the game associated with example (38):

{(Alice snored or John slept, R)}.

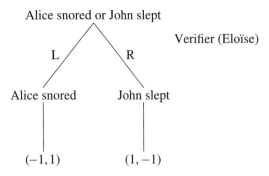

Figure 5.7
Extensive Game Tree for a Simple Disjunction Game

Table 5.5
Material Implication

Sentence₁	Sentence₂	if Sentence₁ then Sentence₂
true	true	true
true	false	false
false	true	true
false	false	true

In the game for example (38), Abélard doesn't get to move at all. The game goes to Eloïse, given the model in table 5.1, so the sentence is true on that model.

Material Implication Logic mavens will recognize that I've already defined a complete system of truth functions (and then some). This means that any possible function on the truth values, "true" and "false," can be defined using the resources I've given. I'd like to give one more definition, simply because it will come up in later chapters. The conditional is illustrated in (39):

(39) If it snows, then I'll stay inside by the fire.

Finding a satisfying interpretation of sentences like the one in (39) is remarkably difficult. I accept the standard logical interpretation of such sentences, material implication (table 5.5), although many people find it a rather pale imitation of a natural language conditional sentence.

The definition in table 5.5 says that a material implication is true if Sentence₁, the antecedent of the conditional, is false or if Sentence₂, the consequent of the conditional, is true. The idea is that this kind of implication should support statements like scientific laws. Consider the following:

(40) If I dropped my uncle from this building, then he would accelerate at 32.2 feet per second per second.

The statement in (40), call it "the law of falling uncles," follows from a physical law governing acceleration due to Earth's gravity. Now, I didn't drop my uncle from the building, but the law remains true even though the antecedent of the conditional is false.

As it happens, though, according to the truth table, table 5.5, the following statement is also true, since I haven't dropped my uncle off any building:

(41) If I dropped my uncle from this building, then he would float up magically and land on the moon.

Of course, the sentence in example (41) is utter nonsense. But since the antecedent is false, the sentence winds up being true (by logic). There have been a number of different approaches to this problem involving possible (but nonactual) worlds or situations.

Jon Barwise tells the story of a poor philosophy student in the 1960s who was arrested at an antiwar demonstration with a rock in his pocket. Charged with carrying a concealed weapon, he pleaded innocent. He was asked, "If someone had attacked you, would you have defended yourself with this rock?" The poor fellow didn't know the answer—he hadn't been attacked, after all—but he did know logic. He reflected on a truth table like table 5.5 and reasoned that since the antecedent of

(42) If someone attacked me, then I would defend myself with this rock.

is false, the whole sentence is true. So he answered yes and was convicted of carrying a concealed weapon. Logic, like other concealed weapons, should be handled with care.

People do occasionally use conditional sentences with a truth table like table 5.5, but it usually marks that the speaker believes that the antecedent is obviously false:

(43) a. If my grandmother had wheels, she'd be a bus.
     b. If John is on time, then I'm a monkey's uncle.

According to table 5.5, the only way for an implication to be false is if the antecedent is true and the consequent is false. The sentences in (43) have the property that their consequents are obviously and outlandishly false. The only way to interpret them as true would be if the antecedents were false. So these examples do seem to follow table 5.5; as always, they carry extra meaning that does not follow from the truth conditions alone; they imply that the antecedent sentences are obviously false, as in example (43a), or that the speaker has reason to suppose that the antecedent is unlikely, as in (43b).

In table 5.5, it is easy to see that the whole sentence should work out as true if either of the following conditions holds:

- The antecedent of the conditional is false.
- The consequent of the conditional is true.

This is expressed in the rule in (44):

(44) (**R.if S then S**)
Suppose the game is $G(S; M)$, and S is

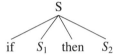

Then the current verifier selects one, and only one, of the following two options:

- The current verifier and the current falsifier exchange roles and the game continues as

$G(S_1; M)$

- The game continues as

$G(S_2; M)$.

That is, the game on a sentence like (45) involves a choice between the negation of the antecedent—the antecedent is taken to be false—and the consequent. Another approach would be to make the verifier first test the antecedent of the conditional and then test the consequent.

Let's examine a game on (45) using the rule in (44):

(45) If Alice snored then Bill walked.

The model in table 5.1 shows that *Alice snored* fails to hold and *Bill walked* also fails to hold. Figure 5.8 shows the game tree. The root of the tree is the sentence in (45). The move is to Eloïse, who must decide whether to play falsifier on *Alice snored* or verifier on *Bill walked*. If she plays falsifier on the antecedent, she wins. Eloïse, then, has a winning strategy on the antecedent in figure 5.8, and the sentence is true:

{(if Alice snored then Bill walked, L)}

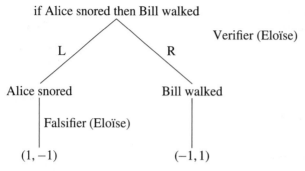

Figure 5.8
Conditional Game

Let's take a case where the condition is false on the model:

(46) If Bill snored then Alice snored.

In example (46), the antecedent of the conditional, *Bill snored*, holds in the model, while the consequent, *Alice snored*, fails to hold. If Eloïse were to pick the antecedent, then she would have to play falsifier and she would lose, since *Bill snored* holds. If she were to pick the consequent, she would again lose because *Alice snored* fails. She therefore has no winning strategy on (46), given the model in table 5.1. The game goes to Abélard, which means that the sentence is false.

I've taken time in this section to show how the logical connectives can be treated as competitive games. The resulting system is equivalent to propositional logic, the logic truth functional connectives. The semantics of the system follows from the model plus the rational choices made by the players when they apply the game rules to the model. It is clear that there is a close and interesting connection between logic and game theory. In the next two sections I explore these connections further by extending the system to first-order logic and then to more exotic logics.

The Aristotelian Square of Opposition
This section discusses the logical operators that play a crucial role in first-order logic, the logic associated with Aristotelian syllogisms. They are *every*, *some/a*, and *no*, as illustrated in (47):

(47) a. Every monkey snored.
　　b. Some monkey snored.
　　c. A monkey snored.
　　d. No monkey snored.

These items have received an enormous amount of attention in both the philosophical literature and the theoretical linguistics literature. In this chapter, I look at these operators from two perspectives. First, there is the traditional truth conditional perspective that says that example (47a) is true if and only if all the objects that count as monkeys in the model also count as snoring things.

The second perspective has received somewhat less attention in the truth conditional approach, but it has a very natural account in the game approach. It has traditionally been assumed that items like *every monkey* fail to refer. The argument goes somewhat as follows. A noun phrase like *every monkey* can be associated with a pronoun within a sentence, as illustrated in example (48):

(48) Every monkey thinks he's clever.

The pronoun *he* in (48) is said to be bound by the noun phrase *every monkey*. This is intended to capture the reading where every monkey thinks of himself that he is clever. That is, each monkey thinks,

(49) I am clever.

Of course, each monkey might believe that all the other monkeys are fools, but as long as each monkey believes of himself that he is clever, the sentence in (48) is true. We can all agree that this is a possible interpretation of example (48).

It is usually held, though, that a quantified expression like *every monkey* cannot bind a pronoun *across* sentences. The following discourse is taken to be defective if the pronoun *he* is taken to be each monkey:

(50) Every monkey chattered. He made a lot of noise.

Once again, I think we can all agree that the discourse in (50) is quite peculiar on the intended interpretation.

It has usually been assumed that the reason the discourse in (50) is odd is because quantified noun phrases don't actually refer. But that can't be quite right either; a quantified noun phrase can establish a discourse object, as shown by the following:

(51) Every monkey chattered. They made a lot of noise.

Notice that *they* in example (51) can be associated with the discourse object established by *every monkey*, that is, the plurality of chattering monkeys. The reading is different from the bound pronoun reading, as shown by the following example:

(52) Every monkey chattered so he made a lot of noise.

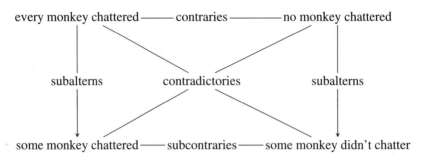

Figure 5.9
Aristotelian Square of Opposition

In example (52) each individual monkey is both chattering and, as a result, making a lot of noise.

Compare the interpretation of (52) with the possible interpretations of the text in (51). In the latter, the first sentence asserts that each and every monkey chattered. But the second sentence is vague. The monkeys collectively made a lot of noise, but the sentence is indeterminate as to whether any individual monkey made a lot of noise. It could be, for example, that each monkey chattered quietly to himself, but the result of all the monkeys chattering was that they collectively made a lot of noise. Some of the monkeys might be chattering quite loudly, and some might be chattering sotto voce; it can't be determined from the text.

Thus, although quantifier noun phrases cannot *bind* pronouns across sentences, they can establish discourse referents that survive through the whole discourse. These discourse elements can be picked out by pronouns and other expressions that act as *discourse anaphora*, that is, they pick out elements that have been established earlier in the discourse. (see part III).

The following methodological principle is important:

Any adequate semantic treatment of an expression must account not only for its truth conditions but also for its impact on discourse.

In other words, one must account not only for the logic of single sentences but also for the logic of conversations.

Let's now turn to the problem of developing games for the Aristotelian determiners. I've chosen to present the data in a form that Peter Abélard would be comfortable with, although logicians today would warn of the shortcomings of the presentation. Figure 5.9 shows the data in the form of an Aristotelian Square of Opposition. The figure marks out some of the important logical relations between the determiners:

1. Two propositions are *contradictory* if and only if they cannot both be true and they cannot both be false.

Thus, *every monkey chattered* cannot simultaneously hold with *some monkey didn't chatter*. One or the other must be true. Equally, *no monkey chattered* and *some monkey chattered* cannot both be true, but one of them must hold.

2. Two propositions are *contraries* if and only if they cannot both be true but can both be false.

Thus, *every monkey chattered* and *no monkey chattered* cannot both be true, but they can both be false. Take the case where there are three monkeys, two of which chattered; *every monkey chattered* is false, and *no monkey chattered* is also false.

3. Two propositions are *subcontraries* if and only if they cannot both be false but can both be true.

Thus, *some monkey chattered* and *some monkey didn't chatter* are subcontraries because they can both be true (see the example given in the previous item, where both are true) but cannot both be false.

4. A proposition is a *subaltern* of another if and only if it must be true if its superaltern is true, and the superaltern must be false if the subaltern is false.

This is really the entailment relation (see chapter 1). Notice that *every monkey chattered* entails *some monkey chattered*; if the former is true, the latter must be true; and if the latter is false, the former must also be false. Equally, *no monkey chattered* entails *some monkey didn't chatter* for parallel reasons.

Existential Sentences Now that the logical structure has been described, let's build some game rules that will account for the Aristotelians.

(53) Some monkey chattered.

Assume, for the moment, that the example in (53) has exactly the same truth conditions and discourse effect as

(54) A monkey chattered.

Both (53) and (54) have a similar effect on the discourse; they introduce a new discourse entity:

(55) a. Some monkey chattered. It was excited by a passing leopard.
     b. A monkey chased John. It was angry at him.

In both (55a) and (55b), the first sentence introduces a new discourse entity (a monkey) and says something about it. In the second sentence, the pronoun *it* is used to refer to the monkey introduced in the previous sentence.

In order for a sentence like (53) or (54) to be true, it must be the case that the verifier can produce a witness, namely, a monkey that chattered. If she can do that, she wins. In addition, the witness that she produces remains available; it is established in the discourse, and the participants can use it later.

To account for this last fact, assume that Eloïse and Abélard have a bag—call it a *discourse model*, or M_D—available to them into which they can put things. Both players will have access to the contents of M_D under certain conditions. We can think of the discourse model as a data structure that the players can use as a resource while playing their games. The game rules regulate how this resource is used.

Here is a version of the rule for *some* and *a*:

(56) **(R.some/a)**

Suppose the game is $G(S; M)$, and S contains an NP of the form

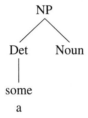

Then the current verifier selects the name of an entity:

such that X has the property named by the original Noun in NP and the game continues as

$G(S'; M)$,

where S' is the result of substituting NP' for NP in S.

In addition, if the current verifier is Eloïse, then the entity named by X is placed in M_D, the discourse model. Otherwise, the *mereological sum* of the entities in the model named by the original Noun in NP is placed in M_D.

The game replaces the logically active noun phrase with *some* or *a* as a determiner with the name of an entity that has the property associated with the noun in the original noun phrase. Thus, a logically active element is eliminated and replaced with a name that picks out an entity that will, if all goes well for the verifier, witness the truth of the sentence.

Mereology is a calculus of parts and wholes; it tells us how to assemble parts to make a larger whole. It is, however, sublimely indifferent to what those parts are. I can take the mereological sum of my left nostril, Notre Dame cathedral in Paris, and my dog Sami's favorite squeaky toy. The result is a perfectly good, albeit somewhat peculiar, whole. The discussion here generally takes the sums of individuals to make pluralities of people and things. For example, a plurality of tigers can be constructed from individual tigers.

Why can't these entities just be accumulated into sets, and the sets used as stand-ins for pluralities? A full argument would sidetrack the discussion, so I just note that pluralities can act together as an individual, as in

(57) The men lifted the piano into the truck.

Indeed, sometimes people act in a coordinated fashion to accomplish something,

(58) The children surrounded the barking dog.

while at other times each individual in the plurality must act alone,

(59) The children slept.

where it is hard to imagine what it means for children to sleep as a group; instead, each member of the plurality sleeps.

It seems that sets don't provide a rich enough structure to account for these differences. Mereological sums offer a variety of model-theoretic objects to model the semantics of these various uses of the plural.

Notice that this requires that all the objects in the model be named. This task is not quite as hard as it seems. Naming could just be a method of pointing to the desired entity and saying "that thing over there." This is what Bertrand Russell called a logically proper name.

Finally, this rule stands in for the existential quantifier in first-order predicate logic. This quantifier is usually written with the symbol $\exists$; for example, *some monkey chattered* would be written

$\exists x[\text{Monkey}(x) \wedge \text{Chatter}(x)]$,

which is read as "there exists an x such that x is a monkey and x chatters." Since the verifier, Eloïse, is the player for existential quantification, she might be renamed $\exists$loïse.

Table 5.6
Model for Example (60)

Expression	Model
Andrew	a
Bert	b
Chrysanthemum	c
tiger	a, c
walked	b, c

The best way to understand the rule in (56) is to see how it applies in a given case with a particular model.

(60) Some tiger walked.

Suppose Eloïse and Abélard are given the model in table 5.6 to use for game play. Once again, we don't need to give a full model of the world, just a submodel appropriate for the game at hand. The model in table 5.6 is sufficient to show that Eloïse has a winning strategy. In this model, there are two tigers, Andrew and Chrysanthemum, and two entities that walked, Bert and Chrysanthemum.

The game is G(some tiger walked; $M_{T5.6}$), where $M_{T5.6}$ is the model in table 5.6. As usual, the model, the sentence, and the game rules jointly determine an extensive game tree. The utilities for *some tiger walked* are the following:

$u_{\text{Abélard}}$(Andrew walked) $= 1$.

$u_{\text{Eloïse}}$(Andrew walked) $= -1$.

$u_{\text{Abélard}}$(Bert walked) $= 1$.

$u_{\text{Eloïse}}$(Bert walked) $= -1$.

$u_{\text{Abélard}}$(Chrysanthemum walked) $= -1$.

$u_{\text{Eloïse}}$(Chrysanthemum walked) $= 1$.

Although it holds in $M_{T5.6}$ that Bert walked, Bert fails to be a tiger, so Eloïse loses on that choice.

The extensive game tree for G(some tiger walked; $M_{T5.6}$ is shown in figure 5.10. Eloïse has a winning strategy:

{(some tiger walked, R)}.

A Game Logic for Natural Language 157

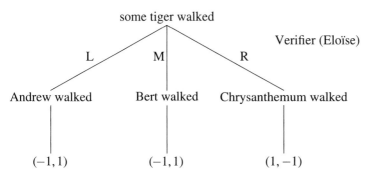

Figure 5.10
Existential Game

As is often the case in such simple games, Abélard doesn't even get to move. Since Eloïse has a winning strategy (play "Chrysanthemum"), the sentence *some tiger walked* is true in model $M_{T5.6}$.

Pronouns When Eloïse selects a tiger to witness *some tiger walked*, that tiger is placed in the discourse model, M_D. The entity that witnessed the sentence is then available to be referred to later in the discourse:

(61) Some tiger walked. She was hunting a monkey.

The pronoun *she* in (61) refers to the tiger that was walking. It's necessary to carefully distinguish the pronoun in (61), which is dependent on an element established earlier in the discourse, from the following kind of example:

(62) Every monkey thinks he's clever.

The pronoun *he* in (62) acts like a bound variable; that is, the pronoun covaries with the monkeys so that each monkey thinks of himself, "I am a clever monkey."

In order to account for discourse pronouns, we need a rule like the following, where $S[X/\text{she}]$ means the result of replacing the pronoun *she* with the name X in the sentence S:

(63) **(R.she) (Discourse Pronoun)**
Suppose the game is $G(S; M)$, and S contains the pronoun *she*. Then the current verifier may choose an entity, X, from the discourse model M_D and replace *she* by the name of that entity. The current falsifier may also choose an entity, Y, from M_D. The game then continues as

$G(S'; M)$,

where S' is

$S[X/\text{she}]$ and X is female and if Y is female then $X = Y$.

The idea behind the rule (**R.she**) is that the verifier can pick someone (or something) as a witness to a sentence containing a pronoun as long as the witness exists in the discourse model; that entity must have been talked about. Furthermore, that entity must be unique in M_D; that is why the falsifier also gets to pick an entity. If that entity is also female, then the verifier will lose. This isn't quite right. It's discussed further in chapter 8.

Consider the following:

(64) Mary was investigating Susan's business. She found some questionable deals.

In this example, there are two candidates for the pronoun *she* in the second sentence: Mary and Susan. The example is completely acceptable, and no one has any trouble interpreting *she* as picking out Mary.

Now look at the texts in (65):

(65) a. It's false that some tiger snored. She was walking.
 b. It's false that some tiger snored. They were wide awake.

The phrase *it's false that* can be taken to be an instruction to the falsifier and the verifier to exchange roles. At this point, Eloïse will play as the falsifier and Abélard will play as the verifier. But Abélard's individual choices are never preserved in the discourse model, M_D. Instead, the *mereological sum* of the entities named by *tiger* would be placed in M_D. This sum just takes all the individual tigers and assembles them into a group, *tigers*. The operation of creating a mereological sum is represented by $\oplus$.

Suppose *it's false that some tiger snored* is true. This means that Eloïse wins on the whole game. But for her to win, it must be the case that Abélard fails to find a witness for *some tiger snored*. That is, he can't find a snoring tiger. It doesn't make sense to say that he can put his witness into M_D. When we get to the continuation sentence *she was walking*, Eloïse won't be able to choose a discourse entity from M_D for the pronoun *she*. This approach seems to predict that the text in (65a) is peculiar; there's no entity in the discourse model for the pronoun to pick out. Now, I think that the text in (65a) does have a valid interpretation, namely, that there is a particular tiger that didn't snore. This may not be the preferred interpretation, but it's a possible one. This example is taken up again in example (81).

Now, since the tigers are cited in $M_{T5.6}$, we can assemble them into

Tigers = Andrew ⊕ Chrysanthemum

and drop them into the discourse model M_D. Since the tigers are now in the discourse model, the verifier has a perfectly good witness for the plural pronoun *they* as in example (65b). The pronoun in that example should just pick out the tigers, which is correct.

Universal Sentences Let's define another corner of the Aristotelian square:

(66) **(R.every)**
Suppose the game is $G(S; M)$, and S contains an NP of the form

every

Then the current falsifier selects the name of an entity

such that X has the property named by the original Noun in NP and the game continues as

$G(S'; M)$,

where S' is the result of substituting NP′ for NP in S.

In addition, the mereological sum of the entities in the model named by the original Noun in NP is placed in M_D.

The rule **(R.every)** is a version of the universal quantifier in first-order predicate logic. This quantifier is written ∀; for example, *every tiger dreamt* would be written

∀x[Tiger(x) → dreamt(x)],

Table 5.7
Model for Example (67)

Expression	Model
Andrew	a
Baxter	b
Chrysanthemum	c
tiger	a, b, c
slept	a, b, c
dreamt	b, c

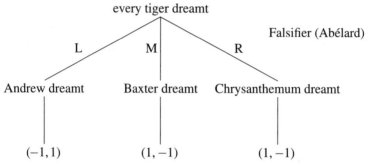

Figure 5.11
Extensive Game Tree for Example 67 Given Model $M_{T5.7}$

which is read as "for every x, if x is a tiger, then x dreamt." Since the rule is a play for the falsifier, usually Abélard, his name might be written as ∀bélard. Thus, our players would be ∃loïse and ∀bélard.

By way of illustration, consider another model, shown in table 5.7. In this model, there are three tigers. All three tigers slept, but only two of them dreamt. The sentence in (67) is relative to model $M_{T5.7}$:

(67) Every tiger dreamt.

According to model $M_{T5.7}$, there is a tiger, Andrew, who fails to dream. Abélard need only play this tiger to win the game. The extensive game tree is shown in figure 5.11. The reader should work out the utilities and confirm that Abélard does indeed have a winning strategy for the game G(every tiger dreamt; $M_{T5.7}$).

By way of contrast, consider what happens on the same model with the sentence

(68) Every tiger slept.

A Game Logic for Natural Language

Inspection of model $M_{T5.7}$ reveals that Abélard has no winning strategy on G(every tiger slept; $M_{T5.7}$). Whichever tiger Abélard picks, that tiger slept, so Abélard will be unable to find a counterexample to the sentence in (68).

Notice that the rule in (66) concludes with the falsifier's placing in the discourse model, M_D, the mereological sum of the entities named by the original Noun in NP. This means that the plurality of tigers in (68) should be available as a target for a pronoun in a later sentence:

(69) Every tiger slept. They were tired from hunting.

The plural pronoun *they* in the second sentence of example (69) should denote the tigers; thus, the second sentence should be interpreted as "the tigers who slept were tired from hunting."

Bound Pronouns Now we can observe a pleasant side effect of the game analysis. Look at the text in (70):

(70) Every tiger thinks he's fierce. They are very vainglorious.

The first sentence can be understood as claiming that every tiger thinks of himself that he's fierce. That is, if the first sentence is true, whichever tiger Abélard picks has the property that he thinks of himself as fierce; the tiger that Abélard picks is thus available to fill in for the pronoun subject of the subsentence *he's fierce*. We need a new rule for bound pronouns:

(71) **(R.he)** (*Bound Pronoun*)
Suppose the game is $G(S; M)$, and S contains the pronoun *he*. Then the current verifier may choose an entity, X, from among the entities previously selected in the course of playing the game and replace *he* by the name of that entity. The current falsifier may also choose an entity, Y, from M_D. The game then continues as

$G(S'; M)$,

where S' is

$S[X/\text{he}]$ and X is male and if Y is male then $X = Y$.

I've simplified things somewhat here; there are syntactic constraints on bound pronouns that I've omitted for ease of presentation. The rule is sufficient to give the interpretation of the first sentence of (70). Once Eloïse and Abélard are finished with play on that sentence, the entity that Abélard selected is lost, but the mereological sum of the tigers remains and is available to stand in for the pronoun in the next sentence.

Table 5.8
Model for Example (72)

Expression	Model
Andrew	a
Bruce	b
Chrysanthemum	c
Xerxes	x
Yolanda	y
Zander	z
tiger	a, b, c
monkey	x, y, z
chased	$\langle a, x \rangle, \langle b, y \rangle, \langle c, z \rangle$

Remember that they must be given a name, say, *those*, a logically proper name in Bertrand Russell's sense.

Scope Phenomena There is another interesting consequence of the game analysis. Consider the sentence in (72):

(72) Every tiger chased a monkey.

The sentence in (72) is ambiguous. One interpretation can be

(73) There is a monkey that every tiger chased.

Example (73) holds in a model if there is a single monkey that has the property that every tiger chased it. The other reading is

(74) Every tiger chased some monkey, not necessarily the same one.

On the latter reading, we simply have to produce, for each tiger, the monkey that tiger chased. There is a model where Eloïse wins on example (74) but loses on (73); this property means that the sentence in (72) is genuinely ambiguous: there is a single model where one interpretation is true and the other interpretation is false.

Actually, the system already accounts for the ambiguity of example (72). The trick lies in the fact that an order of play has not been established for rules like **(R.some/a)** in (56) and **(R.every)** in (66). Let's explore this by working out the example with respect to a model. (The reader can use backward induction to work out winning strategies for these games; see chapter 4.)

Table 5.8 shows a possible model for example (72). Notice that on model $M_{T5.8}$ it's true that every tiger chased a monkey—Andrew chased

Xerxes, Bruce chased Yolanda, and Chrysanthemum chased Zander—but no single monkey has the property that he was chased by every tiger.

Let's now study the game G(every tiger chased a monkey; $M_{T5.8}$). Either **(R.some/a)** or **(R.every)** could apply, so let nature decide by flipping a coin. Suppose it comes up that **(R.every)** goes first; Abélard needs to find a tiger that failed to chase a monkey. He has three choices: Andrew, Bruce, or Chrysanthemum. But for each of these, Eloïse can reply with a monkey that Abélard's tiger chased. Abélard can't win on this version of the game, G(every tiger chased a monkey; $M_{T5.8}$), so the game goes to Eloïse, and the sentence is true on $M_{T5.8}$.

Figure 5.12 shows the extensive game tree for this order of play. No matter what Abélard's choice is, Eloïse has a winning reply:

$\{(L, 1), (M, 2), (R, 3)\}$.

The figure marks Eloïse's winning plays with up-arrows. Since Eloïse has a winning move no matter what Abélard chooses, the sentence is true on the model $M_{T5.8}$.

Of course, nature was kind to Eloïse in forcing Abélard to go first. Suppose the coin toss had selected **(R.some/a)** as the first move. This is the version of *every tiger chased a monkey* that corresponds to the paraphrase in (73). In this case, there must be a single monkey that was chased by every tiger (regardless of what happened to the other monkeys). No one monkey in the model gets chased by every tiger, so Eloïse should lose.

Figure 5.13 shows the extensive game tree for this version of the game. I've marked Abélard's winning choices with a skull and cross bones. Notice that he has a variety of winning strategies; for example, the following profile guarantees that he'll win:

$\{(L, 3), (M, 1), (R, 2)\}$.

He has other strategy profiles that will guarantee a win, but we need only specify one to prove that this interpretation of *every tiger chased a monkey* is false on the model $M_{T5.8}$.

Suppose the model in table 5.8 is modified slightly to the one given in table 5.9. The only difference between the two models is that there is a monkey, Zander, who gets chased by each and every tiger. This change to the model means that Eloïse has a winning strategy on sentence (72) if nature picked her to go first. The extensive game tree is shown in figure 5.14. In this version of the game, Eloïse has a winning strategy if she simply plays R; that is, Eloïse's winning strategy is to pick Zander.

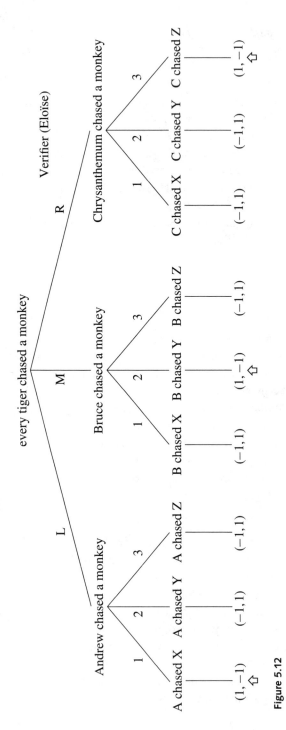

Figure 5.12
A Play of *every tiger chased a monkey*

A Game Logic for Natural Language

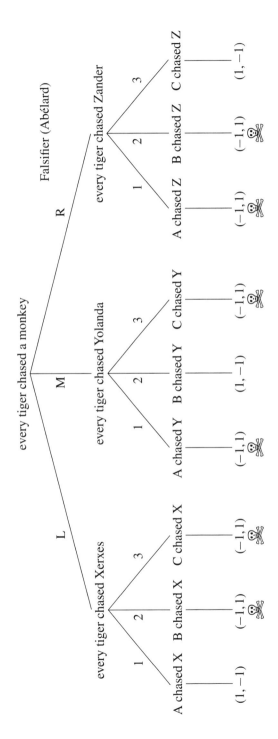

Figure 5.13
Another Play of *every tiger chased a monkey*

Table 5.9
Another Model for Example (72)

Expression	Model
Andrew	a
Bruce	b
Chrysanthemum	c
Xerxes	x
Yolanda	y
Zander	z
tiger	a, b, c
monkey	x, y, z
chased	$\langle a, x \rangle, \langle a, z \rangle, \langle b, y \rangle, \langle b, z \rangle, \langle c, z \rangle$

In this case, there was one monkey, Zander, who had the property that he was chased by every tiger. Eloïse has a winning strategy by just selecting one monkey, so this supports the following kind of text:

(75) Every tiger chased a monkey. He ran up a tree.

The individual that Eloïse chose becomes part of the discourse model. Another way of thinking about it is that the pronoun in (75) could be replaced by a description like the italicized phrase in (76):

(76) *The monkey that every tiger chased* ran up a tree.

Let's compare this case with the other reading associated with example (72), illustrated in the game tree shown in figure 5.13. Recall that in that case Eloïse's strategy was contingent on which tiger Abélard selected. This suggests that the mereological sum of the monkeys in Eloïse's strategy should be placed in the discourse model. We would expect the following text to be valid:

(77) Every tiger chased a monkey. They ran up a tree.

In this case, the pronoun *they* in the second sentence of (77) could be replaced by the italicized phrase in (78):

(78) *The monkeys that the tigers chased* ran up a tree.

In other words, the strategies that the players use can be converted into elements in the discourse model. This results in a connection between strategic behavior and the way that information is used in building a discourse. The needed information is readily available in the verifier's

A Game Logic for Natural Language

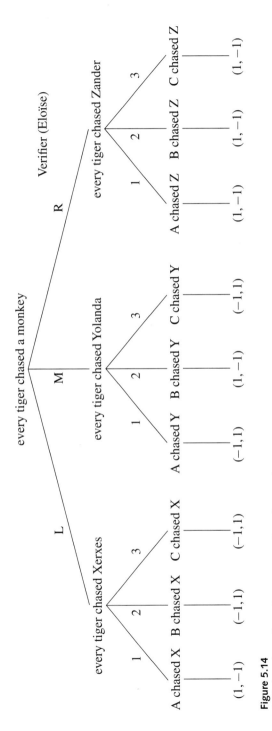

Figure 5.14
Still Another Play of *every tiger chased a monkey*

strategy profile; in this case, Eloïse's strategy profile amounts to the function in (79):

(79) $\begin{bmatrix} \text{Andrew} & \mapsto & \text{Xerxes} \\ \text{Bruce} & \mapsto & \text{Yolanda} \\ \text{Chrysanthemum} & \mapsto & \text{Zander} \end{bmatrix}$

The input to the function is Abélard's potential choice of counterexample, while the output of the function is Eloïse's best response.

If we return to the game shown in figure 5.14, we see that Eloïse's best response is always Zander; this gives a constant function:

(80) $\begin{bmatrix} \text{Andrew} & \mapsto & \text{Zander} \\ \text{Bruce} & \mapsto & \text{Zander} \\ \text{Chrysanthemum} & \mapsto & \text{Zander} \end{bmatrix}$

The functions in (79) and (80) can be derived from Eloïse's strategy; they correspond, in traditional model theory, to *choice functions*, also known as *skolem functions*. The mereological sum of the ranges (that is, the outputs) of the functions are placed in the discourse model to provide the antecedents of discourse anaphora like pronouns.

Scope and Negation   Recall the problem I raised with regard to the text in (65a), repeated here as (81):

(81) It's false that some tiger snored. She was walking.

One interpretation of the first sentence in (81) requires that none of the tigers snored; in this case, the second sentence seems anomalous because the pronoun has nothing to refer to in the discourse model. But notice that there is again an ordering choice between the negation—the *it's false that* part of the sentence—and the rule that interprets *some*. If the latter comes before the former, then Eloïse wins if she can pick a witness tiger say, Chrysanthemum, that doesn't snore. In this case, the tiger becomes available in the discourse model, and the pronoun in the second sentence could refer to that tiger. Thus, we have the paraphrase

(82) It's false that some tiger snored. The tiger that didn't snore was walking.

As with quantifier scope, we can allow chance to make the first move and decide whether (**R.didn't**) should be played before (**R.some**), or vice versa.

Of course, in spoken English the difference between the original reading (no tiger snored) and the other reading (there was a tiger that didn't snore) would be marked by intonation, the amount of stress given to *some* when spoken. Such clues are absent from the kind of artificial language, logical English, being studied here for use with first-order logic.

The Rest of the Square Speaking of first-order logic, let's return to the Aristotelian Square of Opposition in figure 5.9. So far, two corners of the square have been studied, the ones occupied by *every monkey chattered* and *some monkey chattered*. The remaining two corners are occupied by

(83) a. No monkey chattered.
 b. Some monkey didn't chatter.

The example in (83b) already follows from the rules without modification. In this case, the intended reading is the one where chance schedules the quantifier rule **(R.some)** before the rule **(R.didn't)**. In other words, Eloïse picks a witness before she changes roles to become the falsifier: there's a particular monkey who fails to chatter.

Now, when is the sentence in (83a) true? I think it's true when every monkey fails to chatter. That is, it has the same meaning as expressed by

(84) Every monkey didn't chatter.

when chance schedules **(R.every)** before **(R.didn't)**; not a single monkey that chatters can be found in the model. A special rule, **(R.no)**, could be written to express this meaning, but we this is not required. That meaning is already expressed by the current set of rules.

Numbers and Structures What we have so far is an English-like language that is equivalent in expressive power to first-order predicate logic without equality, which means that we can express the Aristotelian Square of Opposition but can't yet express cardinal numbers. In order to get cardinal numbers in first-order logic, we need equality. For example, the expression that means 'at least two dogs barked' would be

$\exists x \exists y [\text{Dog}(x) \wedge \text{Dog}(y) \wedge x \neq y \wedge \text{Bark}(x) \wedge \text{Bark}(y)]$,

which is read literally as "there is an x and there is a y such that x is a dog and y is a dog and x is not the same as y and x barks and y barks." Notice that this allows for more than two dogs that bark to be in the model. Specifying exactly two would require more work, since we would have to say that every thing in that model that barks and is a dog is either identical to x or identical to y.

Why would anyone want to define numbers using only existential quantifiers, universal quantifiers, negation, and identity? The project is to show how arithmetic could follow from logic. Suppose one could show how arithmetic follows just from the basic notions of Aristotelian logic. Then one corner of the complex and sophisticated world of mathematics would have been reduced to the sparse landscape of logic; perhaps other parts of mathematics could be made to follow. The result would be a startling reduction of apparently complex notions to simpler ones.

I think that much the same quest animates a lot of the work in natural language semantics. Natural languages reveal an incredibly complex conceptual world in even the simplest narratives. Consider the following:

(85) As the fire died, John began to wake up.

The sentence in (85) establishes an overlapping relation between two complex events spread over time. The word *as* expresses the relation of temporal overlap, but there are some puzzles. Consider the two constituent events: the fire died, and John began to wake up. How is an event like the fire dying spread over time? When does it begin? Is there a moment of time that represents the exact moment when the fire begins to die? When does it end? Is there a pair of adjacent moments when I could say "John is not waking up" at one moment and "John is waking up" at the next? When does it become true that John is awake? Indeed, I've been talking as though time were made up of a series of discrete moments, but might it not be continuous, so that every moment of time contained more moments of time?

Part of the interest in studying natural language semantics comes in working out these ontological problems. It takes expressions in a natural language like English and shows how to map them onto various mathematical structures, like a discrete or continuous model of time. The ultimate goal of such a project would be to reduce the apparent complexity in natural language expressions to a simple set of structures. That is, one would literally show how the ontological complexity of language could be constructed out of simpler stuff. I think that the game model in this chapter is an interesting way to think about this project.

Of course, one might not be interested in deriving the foundations of mathematics from simple models of first-order logic. A linguist might just suppose that the talk of numbers does not need to rest on the foundation of logical analysis; numbers can be regarded as primitive elements of the universe. If that's correct, we could give the following analysis to "at least n" (n is a cardinal number like 2) as a determiner:

A Game Logic for Natural Language

(86) (**R.at least n**)
Suppose the game is $G(S; M)$, and S contains an NP of the form

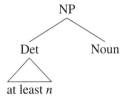

The current verifier selects a set of entities from the model M to act as witnesses for the original sentence. The entities that the current verifier selects must have the property named by Noun. Call this set of entities, **witness**(M). There must be at least n entities in **witness**(M); that is, $|\text{witness}(M) \geq n|$. The current falsifier then selects an entity from **witness**(M) with the name

The game continues as

$G(S'; M)$,

where S' is the result of substituting NP' for NP in S.

In addition, the mereological sum of entities in **witness**(M) is placed in M_D.

The rule in (86) does not attempt to define the number n using logical resources; it simply presupposes that n is defined and understood. I think this is perfectly fair to do, given a lack of interest in the philosophical foundations of mathematics.

As always, the rule ends with the mereological sum of the entities that the verifier has chosen as potential witnesses being placed in the discourse model, M_D, so that they are available to the discourse. This move accounts for texts like the one in (87):

(87) At least four monkeys teased a tiger. They were asking for trouble.

The pronoun *they* in the second sentence in (87) refers back to the monkeys that teased the tiger. Now, I it's clear that if, say, six or eight monkeys teased a tiger, the first sentence in (87) remains true.

The pronoun *they* in (87) should pick out all the monkeys that teased a tiger, but the verifier only has to provide a witness set of four monkeys, not six or eight, in order to verify the first sentence in (87). Somehow it seems distinctly uncooperative of the verifier to leave any monkeys out, but nothing in the semantic rule in (86) requires the verifier to do the additional work of providing extra witnesses.

Here we can see a limitation of the framework developed in this chapter. There is no reason for the verifier and the falsifier to cooperate with each other. But I think that actual communication, whether linguistic or nonlinguistic, involves cooperation. If I know that six monkeys teased a tiger, but I tell you that at least four did, then I've been truthful, but it would be well within your rights to complain that I've been misleading. I knew more than I said, and at very least, that seems uncooperative. You might suspect that I was up to something sneaky by withholding information. Part III considers linguistic analyses using cooperative games where the interests of the players coincide.

Prospects

In this chapter, I have laid out some connections between logic and extensive games of perfect information, and have developed a formal language that looks like a fragment of English but acts, in many ways, like first-order predicate logic. There's a great deal to be said for exploring logic using game theory, and I think the connections are fascinating and useful.

Ambiguity

That being said, it's evident that the correspondence is not perfect. For example, the language of first-order logic is usually defined to be *unambiguous*. That is, first-order logic includes all sorts of notational devices like parentheses that make the structure of each sentence of first-order logic perfectly clear.

The language defined in this chapter, for instance, a sentence like

(88) A tiger chased every monkey.

is ambiguous. The ambiguity in this sentence is resolved by the game tree associated with the sentence. In first-order logic, the sentence in (88) would be associated with two distinct formulas, each of which receives a unique interpretation:

(89) a. $\exists x[\text{Tiger}(x) \land \forall y(\text{Monkey}(y) \rightarrow \text{Chase}(x, y))]$.
 b. $\forall y[\text{Monkey}(y) \rightarrow \exists x(\text{Tiger}(x) \land \text{Chase}(x, y))]$.

So there's a real difference between the process outlined here, which disambiguates an ambiguous language during game play, and regular logic, which disambiguates each sentence by virtue of its syntactic structure.

Monotonicity
There are also many logical relations that haven't been worked out yet. Some, like entailment, have an obvious game-theoretic interpretation. Recall that entailment was defined in chapter 1 as follows:

(90) **Entailment**
A set of sentences $\{S_0, \ldots, S_i\}$ entails another sentence S_m if and only if sentence S_m must be true whenever all the sentences in $\{S_0, \ldots, S_i\}$ are jointly true.

Having defined truth in terms of Eloïse's having a winning strategy, one can give a game version of entailment:

(91) **Game Entailment**
A set of sentences $\{S_0, \ldots, S_i\}$ entails another sentence S_m if and only if in all models where Eloïse has a winning strategy on all the sentences in $\{S_0, \ldots, S_i\}$, she also has a winning strategy on S_m.

Since truth is defined in terms of the verifier's winning strategy, it makes sense to define entailment in terms of winning strategies as well.

Some semantic notions, though, don't have such an obvious game-theoretic treatment. Consider, for example, monotonicity properties. When a quantifier is upward-monotonic or upward-entailing, one is allowed to make an inference from a subset to a superset. For example, priests form a subset of men, which in turn form a subset of the set of humans. The quantifier *some* is upward-entailing:

(92) a. Some priest arrived at the station.
b. Some man arrived at the station.
c. Some human arrived at the station.

Given the truth of (92a), I am justified in concluding (92b) and (92c). The game property involved here seems to be *subgame perfection*. In classical game theory, a strategy profile is a subgame-perfect equilibrium if it represents a Nash equilibrium of every subgame of the original game. This means that if the players played any smaller game that consisted of only one part of the larger game, and their behavior represents a Nash equilibrium of that smaller game, then their behavior is a subgame-perfect equilibrium of the larger game. Subgame perfection usually involves a kind of backward induction from the subgame to the larger game. If the verifier has a winning strategy on an upward-entailing quantifier, then her strategy is subgame-perfect; she has a winning strategy on any game that contains the smaller game.

Not all quantifiers are upward-entailing, of course. *Every* is not:

(93) a. Every priest drank a pint.
  b. Every man drank a pint.
  c. Every human drank a pint.

If I know (93a), I'm not justified in concluding (93b). Although every priest may have lifted a pint, there might be a teetotal man who didn't drink. As it happens, *every* is downward-entailing; if I know (93c), then I'm allowed to conclude both (93b) and (93a). Although this property seems to be related to subgame perfection, it can't be the same.

Notice that many quantifiers are neither upward- nor downward-entailing. For example, *exactly* is neither upward- nor downward-entailing:

(94) a. Exactly four priests danced a jig.
  b. Exactly four men danced a jig.
  c. Exactly four humans danced a jig.

Compositionality

Chapter 1 touched on the idea that natural languages should be compositional:

(95) ***Compositionality***
  The meaning of a phrase is a function of the meanings of its parts and their mode of combination.

The idea is that in working out the meaning of a phrase or sentence, the semantics can follow the syntactic structure that the grammar assigns to it. For example,

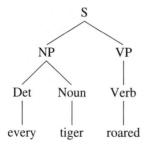

The meaning of the root node, *S*, would be computed as a function of the meanings of the subject NP, *every tiger*, and the meaning of the predicate VP, *roared*, and the way they combine, predication. The meaning of the NP would be a function of the meaning of the determiner *every* and the noun *tiger*.

Compositionality taken in this sense has the consequence that an ambiguous sentence like

(96) Every tiger chased a monkey.

needs to receive two different syntactic analyses, one for each distinct interpretation of the ambiguity.

The game logic developed in this chapter takes a somewhat different approach to computing meaning; it resolves ambiguities in the language by game play. The game logic is not compositional in the sense of (95). Nevertheless, the meaning of a sentence can be computed given the model and the game rules.

Jaakko Hintikka and his students have argued that natural languages are not compositional in the sense of (95). One argument they give is based on sentences like the following:

(97) Every gentleman has a different hobby.

Gentleman 1, say, Percy, has a hobby (say, snail juggling) that is different from all the other gentlemen's, and gentleman 2, say, Reginald, has his own unique hobby (say, moth-collecting), and so on. Various gentlemen could share hobbies—baiting the poor, racing turtles—but each gentleman must have a signature hobby that sets him apart from all the others. Hintikka has shown that this reading cannot be reduced to standard quantifiers. Instead, the meanings of *every* and *a different* have to be put together to form a new game "every-a-different" in a way that makes the intended reading of (97) noncompositional in the sense of (95).

I'm not particularly bothered by the noncompositionality of the game logic here. What most linguists want is a way of computing meaning systematically, and this logic gives that. The game rules plus the principles of zero-sum extensive games of perfect information are sufficient to provide a verification procedure for any sentence in the language. That is, for any sentence in the language, we know what the world must be like in order for Eloïse to verify it or Abélard to falsify it. I can be content with that.

Of course, there may be other reasons why natural languages would adhere to compositionality. Jakub Szymanik has pointed out to me that compositionality might make languages easier for humans to process. Showing this will require joint effort between logicians and psycholinguists.

Limitations

I'm not content with the system presented here as a theory of language, however. I began this chapter complaining about truth-conditional semantics, and I haven't changed my mind about that. Truth conditions might be one aspect of meaning, but they are really only one tiny room in a larger mansion. I would like to break out of the room and see what things look like on the whole estate.

Next, the analysis included a fiction that makes me a little uncomfortable. I suggested that the meanings of lexical items are determinate. Nouns, for example, denote sets of entities, verbs denote either sets of entities or sets of ordered pairs of entities, and so on. Every word has a fixed, determinate meaning that can be expressed in terms of standard set theory.

This is because this kind of truth-conditional semantics needs to simulate a definitional theory of meaning. A definition is intended to provide the necessary and sufficient conditions for the meaning of a word. In model theory, all we have is set theory, so we take the necessary and sufficient conditions to be membership in a set. Thus, *tiger* is among the entities that count as tigers. The theory says that if I know what *tiger* means, I can successfully sort the world into tigers and nontigers. Even poor Claude the tiger (see chapter 1) presents no problems here. He must be a tiger because he's in the extension of *tiger*, meaning that he is an element of the set of tigers.

Now, I don't think this is a particularly good or interesting theory of the conventional meanings of words or phrases. Certainly, there are good reasons to suppose that such a simple theory of word meaning and concepts falls short of providing a satisfactory account of empirical phenomena. As it stands, the theory I've given is an instance of a *correspondence theory* of meaning; each word corresponds to some element of the real world. Of course, things are never that simple. Consider the following example:

(98) The senate voted for cloture.

Does *cloture* denote some element of the real world? It seems as though cloture is less an abstract entity than a sort of strategic move in a larger game.

A Game Logic for Natural Language

Notice that the logic developed in this chapter is deeply linguistic in the sense that complex expressions are reduced to atomic sentences, which either hold or do not hold in the model. It is as though the entire universe were made of sentences. I'm sure that this is how the world works in general—opinions vary—but I think we need to consider how we connect language to the world via usage. In particular, we can afford to be a bit more curious about things like lexical meanings. We need to incorporate some of the insights about the social nature of language into a theory of word meanings.

Another limitation of the game logic is that the games are purely zero-sum; Eloïse and Abélard are locked in a battle over each sentence. But, as I noted with respect to example (87), communication involves cooperative behavior. When I'm talking to you, you assume that I'm being cooperative, telling you the truth, and not leaving things out in order to mislead you.

This means that our interests often coincide; when we converse, we are not playing a zero-sum game. I want to get my meaning across to you, and you want to grasp my meaning. That congruence of interests is the bedrock on which cooperative linguistic behavior rests. These insights should be incorporated into an account of linguistic communication. I argue that, in fact, meaning arises out of the interplay of our common interests and strategic choices. Chapters 7–9 explore this idea. Chapter 6 considers some puzzles about common knowledge and common ground.

Further Reading

I have been charmed by game-theoretic semantics since I first ran across it as an undergraduate acolyte of Wittgenstein. At the time, I was told that the approach was nonstandard, which was enough to scare me off. Now I take "nonstandard" to be a form of praise. I've actually found the game approach to logic to be tremendously helpful. Barwise and Etchemendy (2002) give a good introduction to games and logic.

Games have a long history in philosophy. I can hardly do justice to them here, but see Pietarinen (2003; 2008) and the references cited there for some useful discussion. Hintikka has been developing game-theoretic semantics for decades. A few high points are Hintikka and Kulas (1983; 1985), Hintikka and Sandu (1997), and Hintikka (1996).

The dialogue approach to meaning was, as far as I know, first developed by Carlson (1983). Discussions of game-theoretic approaches to quantifiers can be found in R. Clark (2007; 2009) and Pietarinen (2007). The discussion in this chapter only scratches the surface of the topic.

I decided to give a logic that looked like English after being inspired by an introduction to model-theoretic semantics by Barwise and Etchemendy (1989). I decided to use tree-adjoining grammar (TAG) because it's such a lovely way to treat natural language syntax. Frank (2002) gives a good introduction to TAGs.

Games have become a hot topic in logic over the past few years, particularly with the advent of Hintikka's "independence-friendly" (IF) logic. This logic differs from the logic presented here in that the game is a zero-sum game of incomplete information; thus, instead of the cake game, one plays something like rock-paper-scissors, where the players must make a choice without knowing the other players' decisions. Sevenster (2006) gives a good but rather technical discussion of these kinds of logics. Van Benthem (2011) extends game logics into many areas, including dynamic logic and discourse; this work shows the richness of the game stance for the development of logic and cognitive science.

There is some evidence that IF logics might be appropriate for natural language on the basis of the interpretation of sentences like

(99) Some relative of each villager and some relative of each townsman hate each other.

This is on the basis of the intuition that the choice of relative for each villager is independent of the choice of townsman. This gives me a headache when I think about it too much. See Szymanik (2009) for a sensible discussion.

More recently, Parikh proposed another approach to semantics grounded in game theory, equilibrium semantics, which uses a more sophisticated model theory, situation semantics (Barwise and Perry 1983). See Parikh and Clark (2007) and Parikh (2010) for discussion.

III GAMES AND THE WORLD

6 Common Knowledge

If I want to tell you something, I have to make decisions about *how* to tell it to you. In order for you to figure out what I mean, you need to have a model of *what* I know. Both the speaker and the hearer need to model each other's information state. But how can this work? My model of your information should contain a model of what you think I know. What you think I know includes a model of what you think I think you know. What I think you know will have to include a model of your model of what you think I think you know.

Are we trapped in an infinite regress, a kind of conceptual hall of mirrors? At first glance, it looks bad; we can never finalize our models of each other, since our models would have to contain the models themselves. Barwise and Moss (1996) tried to work out a mathematical theory of non-wellfounded phenomena; these are sets that contain themselves. They are a rich source of logical paradoxes and migraines.

Happily, game theory can ease this non-wellfounded dilemma. The assumption has been that the players in a game all know the game: the choices available to the players, the utilities associated with the outcomes, and so on. The crucial information for modeling is already in the game tree, which is assumed to be public information, known to all the participants. The infinite regress has vanished because the game is, in the relevant sense, public.

The problem is not solved, however. Do speakers and hearers really have perfect knowledge of the game they are playing? The best they can do is *approximate* common knowledge. In principle, true common knowledge can never be achieved. Recall that the analysis in part II followed the standard game-theoretic assumption of rationality. A rational agent is one who has perfect information about the game, has unlimited computational resources, and is self-interested.

This chapter shows that true common knowledge, a prerequisite of rational game playing, is unachievable. But speakers and hearers must achieve something like common knowledge if they are to communicate. This leads to the important notion of *satisficing*, methods of approximating optimal behavior. In support of the hypothesis that interlocutors are satisficing agents, we look briefly at miscommunication and then at *accommodation*, instances where interlocutors repair their store of common knowledge. In the end, speakers and hearers have enough knowledge of the game to allow them to coordinate their linguistic behavior most of the time. As always, the world is more interesting than the simple initial model would suggest.

Coordinated Attack

Two allied armies, the red army and the blue army, are bivouacked on two hilltops, ready to attack. In the valley below them lies the enemy army, the greens. The enemy army is massive, a vast affair with an almost insurmountable advantage in numbers. Even with the geographical advantage of the hilltops, neither allied army alone could hope to defeat the enemy.

But both the blue general and the red general know that if they coordinate a surprise attack in the night, they can exploit their tactical advantages of surprise and geography to engineer a victory. The two generals, then, need only agree on the time of the attack. The blue general sends the red general the following message:

Most Prized Confederate!

Let us together execute a coordinated attack at midnight. I pray you to acknowledge this message!

Humbly,
Blue General

Of course, the blue general needs some acknowledgment of his message. Otherwise—if, say, the messenger were captured or killed—the blue army might attack alone and face certain doom. So the blue general needs to know that the red general received his message.

Later, the red general receives the blue general's message. He sits down to write an acknowledgment, "message received," but then he realizes something: suppose the acknowledgment is lost; the messenger might be intercepted or killed. If that happens, the blue general won't know that the red general has received the message and is planning to join the attack. The red army would then attack alone and get decimated. So the red general decides to ask the blue general to acknowledge his receipt of the acknowledgment. He pens the following message:

O, Blue General!

Message received. Kindly acknowledge!

Warmest Regards,
Red General

and he sends it off to the blue general.

Soon the messages are flying back and forth between the red and the blue generals; each general wants the other to acknowledge his last message, which he fears may have been intercepted by the greens. At each step, neither general is certain that there is an agreement to attack. Common knowledge that they will attack at midnight can never be achieved. Eventually, of course, the green army organizes a sneak attack and defeats the blue and red armies.

While some parts of the story are clearly fictional—when have the greens ever won?—who can deny that the red and the blue have trouble coordinating? The story serves to demonstrate an important lesson:

(1) Common knowledge is a prerequisite for coordination, but common knowledge cannot be achieved in a system with unreliable communication. Coordinated attack, therefore, is not possible in systems with unreliable communication.

The example of coordinated attack has been discussed in the literature on computer networks; in fact, (1) is paraphrased from Fagin et al. (1995). The problem is actually more general and is not simply an artifact of unreliable communication; it involves inherent bounds on knowledge. The coordinated attack problem is closely related to problems in the strategic use of language.

Definite Descriptions and the Mutual Knowledge Paradox

Herb Clark (1992) discusses a problem that he refers to as the mutual knowledge paradox, which seems similar to the coordinated attack problem.

Of interest here is the use and interpretation of *definite descriptions*. For present purposes, definite descriptions are any noun phrase with the word *the* as its determiner. All the following are definite descriptions:

(2) a. The boy with green hair
   b. The woman who is drinking a martini
   c. The queen of England
   d. The best saxophone player in western Tennessee

Definite descriptions are generally taken as denoting unique objects.[1] Their use requires that both the speaker and the hearer can identify the object that the speaker intends to pick out by using a definite description.

The following example of the mutual knowledge paradox is adapted from H. Clark (1992). Suppose that the Prance Theater is having a Festival of Extreme Folk Dancing. Every weekend a new dance troupe is brought in to perform for two nights only. Alice and Buddy have been attending some of the performances.

Let's first consider a simple case. Alice looks online at the schedule for the Prance Theater and notices that this weekend's performance will be by the Hans-Erni Spruengli troupe of Appenzell Sword Dancers, the canton of Appenzell in Switzerland being known for its swordsmanship and dancing. When she sees Buddy, she asks him,

(3) Have you seen the dance troupe performing at the Prance this weekend?

Now, in order for her question to work, she must have reason to suppose that *the dance troupe performing at the Prance this weekend* denotes some particular dance troupe:

(4) Alice knows that *the dance troupe performing at the Prance this weekend* is the Hans-Erni Spruengli troupe of Appenzell Sword Dancers.

But she must also suppose that Buddy will know what that description refers to. Notice that depending on the weekend, the things that *the dance troupe performing at the Prance this weekend* refers to will be different, in which case Alice's question in (3) might misfire. She will have asked a question that is based on a false premise, or presupposition, namely,

(5) Buddy knows that *the dance troupe performing at the Prance this weekend* is the Hans-Erni Spruengli troupe of Appenzell Sword Dancers.

In another scenario, Alice and Buddy check the schedule for the Prance Theater affixed to their refrigerator. They both see that this weekend's performance is by the Hans-Erni Spruengli troupe of Appenzell Sword Dancers. Later, Alice checks the online schedule for the Prance Theater and sees that the Appenzell Sword Dancers have canceled because of an unfortunate rapier mishap in Ypsilanti. That troupe has been replaced by the Hercules Grytpype-Thyne Extreme Morris Dancing Troupe from Wyre Piddle.

Notice that Alice clearly knows *the dance troupe performing at the Prance this weekend* refers to the Hercules Grytpype-Thyne Extreme Morris Dancing Troupe. But she has no reason to suppose that Buddy knows this. She has every reason to believe that Buddy is still in the state described in (5), although now Buddy's knowledge is out of date. Thus we need to require,

(6) Buddy knows that *the dance troupe performing at the Prance this weekend* refers to the Hercules Grytpype-Thyne Extreme Morris Dancing Troupe.

Now suppose that Alice notes the change in the schedule but wanders off to attend to some pressing matter. She sincerely intends to inform Buddy of the change to the schedule but plans to do it later. Alice has left her laptop unattended. Buddy wanders in and sees the browser open to the Prance Theater schedule. Now Buddy knows that *the dance troupe performing at the Prance this weekend* refers to the Hercules Grytpype-Thyne Extreme Morris Dancing Troupe, but Alice doesn't know that he knows it. In other words, we need to require,

(7) Alice knows that Buddy knows that *the dance troupe performing at the Prance this weekend* refers to the Hercules Grytpype-Thyne Extreme Morris Dancing Troupe.

Is this enough?

Now, as before, Alice leaves her laptop unattended and Buddy sees the new schedule. Suppose that Alice sees Buddy reading the new schedule, but he doesn't notice her. Now Alice has the information described in (7). She asks,

(8) Have you seen the dance troupe performing at the Prance this weekend?

She has no reason to suppose that Buddy knows that she knows that he knows that the dance troupe performing at the Prance this weekend is the Hercules Grytpype-Thyne Extreme Morris Dancing Troupe. In fact, Buddy thinks that Alice still thinks that he thinks that the Appenzell Sword Dancers are performing. So, he must be thinking to himself, "I know who's performing, and I know that Alice knows who's performing. But Alice doesn't know that I know. So if she says something about the dance troupe at the Prance, then she probably means those

Appenzellers." If only Buddy knew that Alice knows that he knows that the schedule has changed! Of course, Alice needs to know that Buddy knows all this, so we need to add a requirement:

(9) Alice knows that Buddy knows that Alice knows that Buddy knows that *the dance troupe performing at the Prance this weekend* refers to the Hercules Grytpype-Thyne Extreme Morris Dancing Troupe.

We can extend the story still more. Suppose, as before, Alice leaves her laptop for Buddy to see. She sees Buddy looking at the web page announcing the schedule change from her presumably hidden vantage point, unaware that Buddy can see her reflection in a nearby window and knows perfectly well that she can see him reading the page. Now, Buddy knows that Alice knows that Buddy knows about the schedule change, but Alice doesn't know that Buddy knows that Alice knows that Buddy knows about it. Buddy should suppose, then, that by the definite description *the dance troupe performing at the Prance this weekend*, Alice intends to pick out the Appenzellers and not the Morris Dancers.

Clearly, the problem for Alice and Buddy lies in their achieving a state of mutual knowledge; that is, in order for Alice to use the definite description *the dance troupe performing at the Prance this weekend* felicitously (successfully) with Buddy, she needs to satisfy the following:

(10) a. Alice knows that *the dance troupe performing at the Prance this weekend* refers to the Hercules Grytpype-Thyne Extreme Morris Dancing Troupe.
   b. (Alice knows that Buddy knows that)^m *the dance troupe performing at the Prance this weekend* refers to the Hercules Grytpype-Thyne Extreme Morris Dancing Troupe.

The notation (Alice knows that Buddy knows)^m is interpreted to mean that the items between the parentheses can repeat an arbitrary number of times (including zero times). That is, her felicitous use of the definite description requires that she achieve a kind of fixed point of mutual knowledge; she can infer all the loops of the (Alice knows that Buddy knows)^m system from her current state of knowledge.

Buddy also needs to achieve a fixed point if he is to interpret correctly Alice's use of the definite description *the dance troupe performing at the Prance this weekend*:

(11) a. Buddy knows that *the dance troupe performing at the Prance this weekend* refers to the Hercules Grytpype-Thyne Extreme Morris Dancing Troupe.

b. (Buddy knows that Alice knows that)^n *the dance troupe performing at the Prance this weekend* refers to the Hercules Grytpype-Thyne Extreme Morris Dancing Troupe.

H. Clark (1992) correctly argues that in order for the use of a definite description to be felicitous, the speaker and the hearer must achieve *mutual knowledge*:

(12) A speaker, A, and a hearer, B, *mutually know* that p (p some proposition) if and only if all the following hold:
 (1) A knows that p.
 (1') B knows that p.
 (2) A knows that B knows that p.
 (2') B knows that A knows that p.
 (3) A knows that B knows A knows that p.
 (3') B knows that A knows B knows that p.
 And so on forever.

Now, as Clark observes, the information in (12) can be abbreviated as:

(13) A and B mutually know that p if and only if

 (q) A and B know that p and that q.

But both (12) and (13) involve an infinite recursion.

Common knowledge is fundamental to communication. We usually think of communication in terms of the unknown: if I transmit information to you, then I fill you in on something you didn't know before. While this may be true, I can't convey anything if we don't share some store of knowledge.

The previous example is a case in point. If Buddy doesn't know anything about the Prance Theater, its schedule, the groups performing there, and so on, then Alice has no hope of using *the dance troupe performing at the Prance this weekend* to refer to anything insofar as Buddy is concerned. They need to share a common store of knowledge that they can both draw upon in order to plan strategies for encoding and decoding linguistic forms. This brings us full circle back to a point made in part I: meaning has an irreducible social component because we have to share knowledge of meaning if we hope to communicate with each other. We can't think of linguistic meaning simply as a representation similar to an equation.

On the other hand, we must be careful about how we think about common knowledge. As H. Clark observes, the formulations of mutual knowledge in (12) and (13) seem to rest on an impossibility. In working out whether we have genuine mutual knowledge, we apparently must work through an infinity of statements. But since we are finite beings, we have neither the time nor the memory to verify all the statements.

Another way to think about it is to suppose that in speaking I have reference to a model of your information state. Inside your information state is a model of my information state. Of course, inside your model of my information state is my model of your information state, and so on. We would wind up with recursively nested homunculi like a pair of infinite Russian nested dolls. Actually, there is a branch of set theory—non-wellfounded sets—that studies sets that contain themselves as members.

From the point of view of game theory, there is an obvious resolution for part of the mutual knowledge paradox. Both the speaker and the hearer—Alice and Buddy, in the example—need to be aware of the game they are playing in order for them to play the game rationally (see chapter 4). If both players are aware of the game tree for the communicative game, then the game tree can act as a fixed point. Alice can then reason along the following lines:

> Buddy is a sensible fellow. He's aware of the local context and the linguistic choices that we both have, given this context. I know that he knows this, and he knows that I know it. In fact, we're both aware of the game as if the tree were drawn here in front of us; based on that, I can make any common knowledge inferences I might happen to need. The game itself is common knowledge between Buddy and me.

Buddy has the same kind of reasoning available to him. So, given that both players know the game tree and know that it is common knowledge, we've sidestepped the infinite Russian nested dolls.

Figure 6.1 shows a simple analysis of the game, which suffices for illustrative purposes.

Notice that the game in figure 6.1 is not a zero-sum game. Alice and Buddy are cooperating because they want to communicate. This brings up another important point. The games considered in chapter 5 are strictly competitive, and the players are not engaged in any sort of communication. Eloïse and Abélard verify and falsify statements, given a model. That sort of game is useful for thinking about truth conditions, but we need a cooperative game to study communication.

Now, Alice's ultimate goal is to refer to δ_1—the referent *the Hercules Grytpype-Thyne Extreme Morris Dancing Troupe*—and Buddy's goal is to figure out what Alice is talking about. In figure 6.1, Alice's choices are rooted in a rectangle and Buddy's choices are rooted in circles. Alice needs to select an expression that will communicate her intended meaning

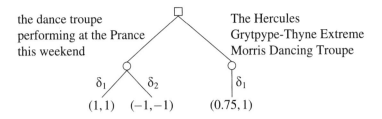

□ = Alice
○ = Buddy
δ_1 = The Hercules Grytpype-Thyne Extreme Morris Dancing Troupe
δ_2 = The Hans-Erni Spruengli Troupe of Appenzell Sword Dancers

Figure 6.1
Simple Communication Game

to Buddy. Her ultimate choice of expression is contingent on Buddy's choice of referent. Buddy needs to select an interpretation for Alice's utterance. What could she mean by what she says?

The hypothesis is that speakers and hearers make the choices they do because they are rational agents who are interested in maximizing their expected utility. The simplified game in figure 6.1 allocates utility as follows:

• Successful communication is preferred to failure.
• Items involving undue production costs—loads of memory or planning—are dispreferred by the speaker.

I'll amend these factors in chapters 7 and 8. For the moment, we have a good starting point for thinking about the factors that condition language use and strategic decision making.

The system outlined here is not a cheap-talk system. A cheap-talk system assumes that speech costs nothing to produce. I advocate a different kind of system. Speech does cost effort—it must be planned, memory is loaded for production and perception, and other resources (time and so on) are used. Although these costs are not monetary, we can still use the preferences that these costs generate to study linguistic decision making. Here I completely agree with Bourdieu (1991); all language involves production costs as well as power relations; language is a full-fledged economic system.

Returning to the game tree in figure 6.1, we see that if Alice says *the dance troupe performing at the Prance this weekend*, Buddy has two choices. He can select δ_1, the intended referent of Alice's utterance, or δ_2, which is the (canceled) Appenzell Sword Dancers.

If Buddy chooses the intended referent, then he and Alice have communicated successfully and both get a payoff of 1. If he selects δ_2, then communication has failed, and both receive a penalty, -1.

Of course, Buddy would choose δ_1 or δ_2 depending on his estimate of how likely it is that Alice meant either group. If he doesn't know that the Appenzell Sword Dancers were canceled and replaced by the Morris Dancers, then his estimate of the probability of δ_1 is 0 (he might not even know about them), and his estimate of the probability of δ_2 is 1, since he believes the (now obsolete) schedule he saw. Alice would be well-advised to say *the Hercules Grytpype-Thyne Extreme Morris Dancing Troupe*, since she gets a positive payoff by doing so; using the definite description *the dance troupe performing at the Prance this weekend* will net both players a negative payoff.

If Buddy knows that the Appenzell Sword Dancers have been replaced by the Morris Dancers (and he has reason to believe that Alice is playing the same game), then his estimate of the probability for δ_1 is now 1, and his estimate of the probability for δ_2 is 0. Notice that the game tree shown in figure 6.1 does not account for this; chapter 7 discusses the right kind of game for this kind of communicative situation.

Suppose that each player is uncertain about the other player's knowledge state. Alice isn't sure what Buddy knows, and Buddy isn't sure what Alice knows. Then Buddy has a fifty-fifty chance of choosing correctly between δ_1 and δ_2, and the expected utility for the players if Alice uses the definite description *the dance troupe performing at the Prance this weekend* is 0. In this case, the choice *the Hercules Grytpype-Thyne Extreme Morris Dancing Troupe* clearly dominates, since it guarantees a positive payoff for both players.

Notice, in figure 6.1, that I've docked Alice 0.25 for choosing *the Hercules Grytpype-Thyne Extreme Morris Dancing Troupe*, the reason being that this utterance is hard to remember. Nevertheless, this may be her best option if her expected utility for choosing *the dance troupe performing at the Prance this weekend* is 0 or -1. This shows that while the interests of the players are not completely at odds, as they would be in a zero-sum game, neither do they coincide perfectly.

So the usual game-theoretic assumptions about common knowledge—that knowledge of the game is shared by the players—can solve part of the mutual knowledge paradox. An infinite recursion of homunculi is not needed to solve the problem. But this addresses only half of the paradox.

Recall the coordinated attack problem. The red general and the blue general could never coordinate their attack on the green army because they couldn't arrive at a state of common knowledge. Try as they might, they were forever tortured by doubt about the other's knowledge state. Indeed, that story provides an informal inductive proof that common knowledge is forever unattainable. The game tree representation is perfectly good as a compact representation, but it is of little help if we can never arrive at a state of common knowledge. How are we to communicate at all, then?

Common Knowledge and Bounded Rationality

When confronted with an impossibility, the sensible thing to do is to admit that it's impossible. While common knowledge is necessary for communication, we have convincing proof that true common knowledge is impossible to achieve. It follows, then, that we are never in a state of true common knowledge, even when we believe ourselves to be. Instead, we act on the compelling illusion that we have true common knowledge with our fellow interlocutors.

We are now in the domain of what has been called *bounded rationality*, following the classic work of Herb Simon. Chapter 4 established an idealization of what constituted rational choice:

- A set A of actions from which the decision maker selects
- A set C of consequences to these actions
- A consequence function g that relates consequences to actions
- A preference relation on the consequences

This model provided the basis for developing games and game trees.

This model, however, assumes that the rational agents are fully aware of the actions available to them, the consequences of these actions, and the relation between actions and consequences. Finally, the agents have a complete preference relation on the consequences of their actions. So the players have almost godlike powers of apprehension, as Eloïse and Abelard do, in their logical heaven.

Back here on earth, however, things are not so easy. I may not be fully cognizant of my actions or their consequences, and I may not have a full grasp of my preferences. I am a *bounded* agent. My capacity to work out the game is limited by time and memory. I have only finite resources to work with. Given all this, I do what I can. I am, to use Simon's term, a *satisficer*. I make do with what I have, even if that means that my choices are not necessarily optimal. Satisficing is doing the best one can with what one has.

Recall the dirty frat boy problem (see chapter 3). When there are n dirty frat boys, any one among them will see $n - 1$ dirty frat boys. The solution to the puzzle relies on this fact; each dirty frat boy reasons from this fact to the theorem that it will take n rounds of questioning by the dean to reveal whether he is, in fact, a dirty frat boy. On the basis of my experience presenting this problem, I suspect that most readers did not immediately see this solution. The reasoning involved in the dirty frat boy problem requires induction on knowledge states and is far from obvious for most people. One has to be taught to see the solution; that indicates the bounds of the usual human reasoning processes.

The rational man of economic theory always maximizes his utility. As a less-than-rational man, I stumble along until I hit on a solution that seems good enough; I satisfice. *Satisficing* is coined from *satisfactory*. The idea is that a satisficer has a subset of actions that are satisfactory; while the results may not be optimal, a satisficer is willing to accept the payoffs as good enough. In a satisficing procedure, the agent tests each action according to some order that has been placed on the actions; she keeps going until one of the actions is in the satisfactory subset, then stops. The result may not be the best possible, but it is good enough, and it saves the satisficer from a more expensive search.

Equally, there might be some simple bound on the computation that the agent is willing to perform. In order to illustrate this, let's take an example of a game that has been tested empirically by behavioral game theorists. It's called a beauty contest, since beauty contests are based on selecting the contestant that most people would regard as beautiful.

Suppose I stipulate that I want everyone in some group to pick a number between 0 and 100. Now take the average of the numbers that people picked. I'll pay ten dollars to the one who picked a number that is closest to 70 percent of the average. What number would you pick if you were in the group?

This game involves social reasoning because each player must base his or her answer on a prediction about how the other players will behave. A good game theorist might reason as follows:

• Since the range of numbers is 0–100, a likely average of the group's choices is 50, which is right in the middle of the range. Seventy percent of 50 is 35.

- But everyone knows this, so everyone will pick 35, which will then be the likely average of the group's choices. Seventy percent of 35 is about 25.
- But everyone knows that, too, so everyone will pick 25, which will then be the likely average of the group's choices. Seventy percent of 25 is about 18.

Thus, hoping to outwit the others, the player chooses 18 as her final guess.

In classical game theory, she would, of course, select the Nash equilibrium by finding the number x^* such that

$$x^* = 0.7 \times x^*,$$

which happens to be zero. The game is dominance-solvable because it can be solved by the iterated calculation of dominance until one lands on a steady state answer. Stepping through the previous exercise shows that the answers approach zero.

Of course, virtually no one picks zero in this game, despite the recommendation of classical game theory. Camerer (2003) reports that most people either pick 35 (they go through one step of the reasoning) or 25 (they iterate twice). Most people tend to compute dominance as bounded to the number of iterations and then stop before the true equilibrium is reached. As people have more experience with the game, they are willing to iterate a greater number of times. Nevertheless, a rational player picking the Nash equilibrium strategy will lose when playing with naive players. In other words, the most rational choice may not be so rational after all.

Let's apply bounded rationality to the problem at hand. What has just been demonstrated is that true common knowledge can never be achieved. Instead, we rely on approximations of common knowledge. These approximations are good enough for most ordinary interactions, although under some conditions things break down, as they did for Alice and Buddy.

In fact, we need a variety of techniques to cover the instances where people assume that they share common knowledge. The problems for Alice and Buddy arose because one or the other agent could not reliably assume that the other had perceived some crucial piece of information. For example, Alice couldn't be sure that Buddy had looked at the laptop, or Buddy couldn't be sure that Alice would know that he had the crucial piece of information.

In many cases, though, the participants can just assume that some piece of information is *self-evident*. A property or event is self-evident for two agents if it can be assumed to be in the information states of the two agents. For example, an object may be physically present in the environment. This idea is adapted from a formal definition by Rubinstein (1998). The basic idea is quite intuitive, although it's mathematically a bit slippery.

The idea can be illustrated with an example. Suppose you and I are sitting at a table on which there is a single glass. I say to you,

(14) The glass is cracked.

This contains a definite description: *the glass*. As with the Alice and Buddy example, use of the definite description involves my expectation that you will be able to work out what it refers to. Since the glass is present perceptually to us both, it seems safe to denote it with *the glass*, particularly since it has been stipulated that it's the only glass on the table.

Further, if our lines of sight are to the same glass, I may assume that the referent is self-evident and use *the glass* to pick it out. Suppose the table has several glasses, but I infer from the direction of your gaze that we are attending to the same glass; then I can use *the glass* even though the glass is not unique.

But the assumption of self-evidence is not the same as common knowledge. However much I might think that the presence of the glass is self-evident, I might be wrong. It might be that you're attending to another glass that I haven't noticed; or that you're lost in thought and haven't noticed the glass; or that I've simply imagined the glass. What is self-evident to me is not self-evident to you.

The requirement for self-evidence is perhaps too broad to be of much use. H. Clark (1996) provides a typology of the possible sources of common knowledge (see figure 6.2). We might take his categories to be a useful way of working out self-evidence.

One source of common knowledge is simply the general knowledge one assumes by virtue of being in a community. By virtue of being an American citizen, I will likely take *the president* to mean the current president of the United States. Someone who is aware of my citizenship might then evoke community membership as a plausible source of self-evidence.

If, however, I'm attending a faculty senate meeting, use of *the president* might be taken as meaning the president of the University of Pennsylvania. This illustrates both that community membership can be taken as indicative of self-evidence and that use of a phrase in a context is crucial in determining its content.

Community membership, however, is not always adequate to establish common knowledge in the sense required by a language game. Although I'm a citizen of the United States, I might not pick up on *the president* as referring to the correct individual. Even at a faculty senate meeting, my mind might be elsewhere so use of *the president* might fail to get me to pick out the president of the University of Pennsylvania.

The example of the glass is an instance of another of H. Clark's plausible sources, *physical copresence*. In the case of the cracked glass, the object is immediately physically copresent; the cracked glass is currently before us, physically copresent with us. In other cases, the object or event may have been copresent with us at a prior time; depending on the salience of the object and the context, we might refer to it.

For example, suppose I put on my kitchen table a form I need to fill out with my wife. I turn away, and a breeze blows it off the table. I might turn to my wife and ask,

(15) Where did *the form* go?

I have made the assumption that my wife was aware of the form and that it was salient to her, as it was to me. In this instance, the copresence of the form was immediately prior to my reference to it, but prior physical copresence is not a sufficient condition for self-evidence. My wife may not have noticed that I put the form on the table, in which case she would be justified in asking *What form?*, indicating that my attempt at reference has failed.

Next, I might use potential physical copresence as the basis for the assumption of self-evidence. This might be a bit harder to see, but consider a case where I have a justified expectation that something will be copresent in the near future. You and I might be watching a film I've seen before, and I say to you,

(16) *The bad guy* shows up in *the next scene*.

In this case, there is a double instance of potential copresence (although this is perhaps a case of virtual copresence); *the bad guy* and *the next scene* are both cases of potential copresence, since we haven't seen either the bad guy or the next scene yet. And, of course, both phrases might fail under certain circumstances. Perhaps I have misremembered the film, or a scene is missing.

While physical copresence is one way that something might be considered self-evident, what H. Clark calls linguistic copresence is another. *Linguistic copresence* means that an item has been mentioned in the discourse already. Consider a sentence like the following as part of a larger story:

(17) John was watching a man in Rittenhouse Square.

The sentence in (17) contains two objects, *John* and *Rittenhouse Square*, which are assumed to be known to the reader (or listener). It also introduces a new object, *a man*. All three objects are now *linguistically copresent*. Another way to think of this is in terms of the discourse models discussed in chapter 5. We could continue the story in a variety of ways:

(18) a. He (= John) is an incorrigible snoop.
    b. The man was acting suspiciously.
    c. It (= Rittenhouse Square) is a great place to watch people in nice weather.

The sentences in (18) show a variety of ways that the story could continue. Each object established in (17) could be a possible topic of the continuation of the story. In examples (18a) and (18c), pronouns are used to pick out *John* and *Rittenhouse Square*. In example (18b), a definite description is used to invoke an entity that was introduced in (17) but that is insufficiently salient to support a pronoun; *John*, the subject of the sentence in (17), is more salient and thus harnesses the pronoun.

The role of salience is crucial. Pronouns can be used to denote things that are highly salient. Schelling's (1960) notion of a *focal point* (sometimes called a *Schelling point*) is relevant here. These are points in the environment that are sufficiently salient as to provide a pinion around which the attention of the various agents in the game can coordinate.

Chapter 9 discusses focal points, and chapter 8, pronouns. For the moment, I summarize some of their properties. Coordination is crucial in defining focal points. Suppose you are visiting New York City with a friend, and the two of you get separated. What do you do? Many people say they would meet near the clock at Grand Central Station or some other well-known location. In Philadelphia, people tend to meet at the Clothes Pin statue by Claes Oldenburg near City Hall or at the eagle in the Wanamaker Building. All of these are highly salient points that can be used to coordinate behavior without the benefit of explicit agreements. In language, things like grammatical function can make an element a focal point with respect to discourse processes like pronominalization.

Both pronouns and definite descriptions, then, assume common knowledge on the part of the speaker and hearer. By virtue of linguistic copresence, the following equations are self-evident to the speaker and the hearer:

he = John.
the man = the man that John was watching in Rittenhouse Square.
it = Rittenhouse Square.

Thus, both the speaker and the hearer assume the equivalences are part of the common knowledge that constitutes the game. This assumption of common knowledge is not fully justified. It is not certain that the hearer didn't find *the man* introduced in (17) more salient than *John* and thus a better focal point. If that were the case, then misunderstandings might be introduced into the game.

H. Clark also comments on a phenomenon that he classifies as *potential linguistic copresence*. This involves elements whose presence in the game can be inferred from the linguistic copresence of other elements. His example is

(19) Because it was broken, I returned the plate I had just bought to the store.

In this case, the pronoun *it* is intelligible because of the later use of *the plate*. One could equally well say,

(20) Because the plate I had just bought was broken, I returned it to the store.

The pairing of (19) and (20) immediately suggests another level of strategic decision making, one more allied to the general problem of stylistics and rhetoric. I leave this as an open problem for the reader.

These examples show that linguistic copresence is an important factor in self-evidence, the property that underlies the assumption of common knowledge between the speaker and the hearer. H. Clark adds still another source of self-evidence, what he calls indirect copresence. *Indirect copresence* is related to our ability to draw inferences from information we already have. For example, imagine a story that begins,

(21) A man walked into a bar.

The sentence in (21) establishes two entities: a man and a bar. The story could continue with

(22) a. The bartender asked him what he wanted.
     b. The mixed drinks cost two dollars because it was happy hour.
     c. The tables were all taken.

Each of the possible continuations in (22) uses definite descriptions that were not explicitly introduced in (21). In each of the continuations, the definite description is something that a reasonable person could infer is

Sources of common knowledge (H. Clark 1996):
1. Community membership
2. Physical co-presence
   a. Immediate
   b. Potential
   c. Prior
3. Linguistic co-presence
   a. Potential
   b. Prior
4. Indirect co-presence
   a. Physical
   b. Linguistic

Figure 6.2
Sources of Common Knowledge

related to a bar. Bars have bartenders, mixed drinks, and tables, among other things. A bar without mixed drinks is a poor one, as is a bar without tables. A bar without a bartender is not a bar at all. Because the bar was mentioned in (21), it was able to bring its canonical properties with it into the discourse model. The speaker can assume that the presence of these things is self-evident to the hearer and that therefore these things are part of the common knowledge she shares with the hearer.

Indirect copresence can also be part of the physical world. An auto mechanic, looking at a car, might say,

(23) The camshaft is frozen.

The presence of the camshaft in the car is inferred from the presence of the car and the fact that all cars have them. In this case, the mechanic assumes (wrongly for someone like me) that the camshaft is self-evident and thus part of our common knowledge.

Figure 6.2 summarizes H. Clark's (1996) list of potential sources for common knowledge. While he includes a more detailed discussion of the possible justifications for these sources, the present list should illustrate the main point: common knowledge is not a given. Instead, linguistic agents are satisficers who use self-evidence to approximate common knowledge.

The theory here, then, is that linguistic agents—speakers and hearers—use a game tree to coordinate signals and messages. Since the game tree cannot be constructed from pure common knowledge, the agents approximate it by using self-evidence. Self-evidence is not a perfect reflection of common knowledge; it is only an approximation, and sometimes it fails so that the agents cannot coordinate their behavior. The game tree is constructed from the agents' awareness of the local model—objects in the discourse model plus a fragment of the model of the world—and whatever options the grammar makes available to them for the purpose of signaling meaning. The local model itself is just the fragment of the world that the agents consider self-evident according to the sources listed in figure 6.2.

The overall behavior of game tree construction in a language game is an example of bounded rationality. The agents satisfice in the sense that they approximate something that they cannot achieve, namely, common knowledge, using the means at their disposal, the sources of self-evidence.

Miscommunication

The bounded nature of game tree construction implies that linguistic agents will sometimes fail to approximate common knowledge, in which case some kind of miscommunication is likely. During the 2008 presidential election, I heard a television commentator say the following:

(24) Colin Powell endorsed Barack Obama because he's black.

A question immediately comes up: Did the commentator mean that Colin Powell endorsed Barack Obama because he, Powell, is black or because he, Obama, is black? The context permitted either interpretation, and I didn't hear anything later that would tell what was intended. Whatever interpretation I might have chosen, I had a 50 percent chance of being wrong, and nothing would have revealed to me that I had erred.

Let's look at a kind of sentence discussed in chapter 5. Quantified noun phrases could interact with each other, resulting in sentences that are truth-conditionally ambiguous (see the discussion of scope phenomena in chapter 5). Suppose I describe to you a class that I taught; I say,

(25) Every student read a book.

On one interpretation, every student read at least one book, possible a different book than all the others read. On another interpretation, all the students read the same book. Which interpretation did I mean in (25)?

The difference in interpretation will affect the inferences that can be made from what I said. For instance, I might continue,

(26) They thought it was poorly written.

The use of the singular pronoun, *it*, reveals that I meant the students all read the same book. On the other hand, I might continue,

(27) They gave presentations on *them* in class.

The plural pronoun *them* indicates that I probably meant that the students read different books.

In the case of (26) and (27), I said something that had the result of revealing which interpretation I intended when uttering (25). A certain amount of uncertainty can be avoided by choosing a different, perhaps more complex, utterance, for example,

(28) a. There was a book that every student read.
     b. Every student read a book, but not necessarily the same one.

If it is crucial that you understand, I might decide to place more effort into making myself clear.

Equally, you might not decide between the interpretations available in (25); you might carry along both possible interpretations, or you might just have an interpretation that is underspecified with respect to the scope relation between *every student* and *a book*.

In any event, there is a chance, when I utter a sentence like (25), that we might misunderstand each other. After that point, our ability to coordinate our game trees could be degraded. The bounded nature of our understanding could change the possibility of our having common knowledge. In fact, we often face ambiguities of a variety of types, any one of which might result in a divergence of understanding that may or may not be revealed by the subsequent context. A misunderstanding could be revealed, in fact, if subsequent discussion reveals that the participants can no longer synchronize their game trees, since their local discourse models have diverged. It's difficult to estimate how often this happens, however.

Since many misunderstandings may not be revealed by the context, we could look to phonological misunderstandings to provide the empirical basis for a working model of misunderstandings. These are cases—it would seem, quite common—where the speaker says one thing and the hearer understands something quite different. A few weeks ago, my wife said to me, with great emphasis:

(29) *Jennifer* I have made a determination.
     I What's May Day determination?

In my defense, the conversation took place toward the end of April, so I thought I heard "May Day determination," which I speculated was some sort of strengthening of the will due to the impending arrival of the first of May.

My mistake involved segmenting the speech stream. Spoken language is not a string of discrete units but a continuous stream of sound. The hearer has to break that continuous stream down into a sequence of discrete parts. In my case, the problem involved the placement of word boundaries, that is, I needed to segment the speech stream into a series of words (the symbol # indicates word boundaries). The decision I faced was between

(30) a. #mejd#ej#dətɚmınejʃən
      'made a determination'
  b. #mej#dej#dətɚmınejʃən
      'May Day determination'

The problem could be represented as a decision tree, where I must choose between the alternatives in (30). Perhaps because I was primed by the arrival of spring, I incorrectly chose *May Day determination*. My confusion revealed that it was a bad choice. If you repeat the phrases *made a determination* and *May Day determination* over and over, I think you'll find there are differences in vowel quality between the indefinite article *a* and the vowel in *May*, even in speech where the indefinite article is pronounced to rhyme with *May*. Nevertheless, when imposing a segmentation on fluent speech, all these factors must be weighed and judged to yield a decision about what the segmentation should be.

Labov (2010) and his colleagues collected a sample of what he calls natural misunderstandings. As with the example in (29), these are cases where the speaker utters one thing, but the hearer understands something else. Say the following sequence of words aloud, concentrating on the position of your tongue and lips during the vowel sounds:

(31) beat, bit, bait, bet, bat.

The different vowel sounds are made by changing the shape of your vocal tract using your tongue, lips, and so forth. You tongue is a kind of fleshy manifold, and its shape and position can alter the resonating characteristics of your vocal tract, which is what gives each vowel its unique characteristics.

When you say "beat," your tongue is high and forward in the mouth. When you say "bit," your tongue relaxes slightly, lowering and changing the acoustic properties of your vocal tract. In all the words in (31), the tongue is forward in the mouth. The difference between them is the height of the tongue; it lowers in each successive word. You can hear the acoustic properties of the vowels more clearly if you shape your mouth as though you were making the vowel and hit your Adam's apple by flicking it with your finger or striking it with a pencil. It hurts, but you'll hear the one of the major resonating frequencies of your vocal tract. As you go through the vowels in (31), you should hear the frequency stepping up.

Crucially, your mouth is not some sort of discrete digital instrument. Your tongue can move around in your mouth continuously. The various vowel sounds are produced by positioning your tongue in this continuous space. You can, for present purposes, think of the vowels in your particular dialect of English as corresponding to ideal positions of your tongue and lips in this continuous space. As you produce speech, you move through approximations of these ideal positions, approaching them as best you can while still producing continuous, fluent speech.

As a listener, I must attend to your speech and make a series of decisions about what vowel sounds you're producing. Notice that this is again a kind of game. You pronounce a word, for example, and I try to select what word you intended based on my understanding of what you produced. Since our vowel spaces may differ, the decision problem is not a trivial one.

It is not clear that speech perception is a game; it might be a pure decision problem for the hearer. That is, given an input, the hearer must decide which vowel sound she has heard. Notice, though, that her decision is based on what she knows about the speaker; that is, she may adjust how she interprets a sound based on knowledge about the speaker. This gives the problem something of the character of a game, particularly if the speaker is also adjusting his output in light of the hearer's behavior.

As Labov and his colleagues have documented, there are a number of distinct sound changes currently happening in American English. Furthermore, these sound changes are affecting different dialects. For example, a distinct series of changes, called the Northern Cities Shift, is affecting cities like Detroit and Chicago, and a Southern Shift is affecting the South. In addition, Philadelphia, which Labov has studied closely for several decades, is showing a number of distinctive changes in its vowel space.

All this suggests that speakers from distinct dialect areas will be prone to misunderstandings. A speaker from one area will have a set of conditioning factors influencing her decisions about vowels that differs from the conditioning factors of a hearer from another area. Labov and his colleagues collected a sample of such misunderstandings, carefully noting the dialect areas of the speaker and the hearer. Here are some examples:

(32) *Telephone surveyor [Chicago]*   Do you have any pets in the house?
    Brian T. [Eastern US] ⇒ pots

In the Northern Cities Shift, the vowel of "pets" is lowering toward the vowel of "pots" in the hearer's dialect. Since the vowels overlap, the hearer has trouble making the decision as to which vowel he actually heard. Another example comes from the Southern Shift:

(33) *Kevin H. [Crossville, AL]*   We have no right ...
    Christina J. [Atlanta] ⇒ We have no rat ...

The following example shows a misunderstanding between two Philadelphians:

(34) *Instructor*   Tell me what this sentence implies to you: "Mr. Williams strode into the office."
    Student   It means he was real casual.
    Instructor   For *strode*? as in *stride*? Do you know what *stride* means?
    Student   I'm sorry, I thought you said "strolled." *Strode* means 'walked forcefully'.

In collecting the misunderstandings, it was noted how a misunderstanding was detected. Labov reports the following:

During the utterance	107
By an immediate query	370
By inference after	199
From observation of later events	72
Never	15

It took fourteen years to collect 869 observations, which works out to a little more than one misunderstanding per week. This might seem low, but it is quite effortful to monitor and record such things, so this is likely an underestimate of actual misunderstandings.

These examples are useful because they show cases where misunderstandings can be detected relatively easily. The discussion with respect to example (25) reveals that some misunderstandings are more insidious. If we assume that semantic and pragmatic divergences happen at least as frequently as the phonological cases but are less likely to be detected, then it becomes apparent that speakers and hearers can fail to coordinate and that satisficed common knowledge might fail. It might be the case that speakers and hearers think incorrectly that their game trees are in synch without ever realizing that communication has failed.

This suggests that the notion of *successful communication* is at best an approximate one. We can think of meaning as a landscape. Using language, I'm able to pinpoint some aspect of that landscape, but it is enormously difficult to reproduce that landscape exactly for you; I can, however, point you to the right area. If you are able to approximate my meaning, then I've done well. The utility of playing a language game cannot depend entirely on exact transmission of information. Instead, we should allow for interlocutors to approach each other's meanings without requiring identity of information. (See chapter 9).

Finally, these cases show the difficulty of coordinating even on such "automatic" tasks as identifying speech sounds, something we have a vast amount of experience with. Thus, despite our expertise, our ability to coordinate our knowledge is bounded and thus prone to the occasional error.

Presuppositions and Accommodation

Communication is not just a matter of what is said but also what is not said. Often, our ability to orient ourselves in the meaning landscape requires us to bring what is not said to bear in such a way that the unsaid becomes part of the act of communication. The preceding section explained how a failure in common knowledge can undermine communication and noted that such failures might even go unnoticed by the participants in a conversation.

This section turns to a case where a failure in common knowledge is not only detected but can be used as a *backchannel* method of communicating additional information. Let's take a famous example. Suppose an attorney is interrogating a witness during a trial and asks,

(35) Have you stopped beating your wife?

Something rather sneaky has happened. In order for the attorney's question to make any sense, it must be the case that the witness at some point beat his wife. Suppose the witness answers, "Yes." Then, he's admitted that he was a wife-beater, albeit a (putatively) reformed one. Suppose he answers, "I haven't." Then the jury is permitted to infer that not only is he a wife-beater but he continues to beat his wife.

There are several things to emphasize about this kind of example. First, whether the witness answers yes or no to the question, the proposition that he beat or beats his wife is assumed to be true. The best course for the witness is to neither affirm nor deny but to take a third course by pointing out that the question is nonsensical.

Second, if the witness answers the question with either a yes or a no, a jury member is permitted to infer that the witness either was or is a wife-beater. The underlying proposition has been communicated, albeit indirectly. If the juror didn't know it before, she knows it now; this process is called accommodation.

In other words, the question *presupposes* that the witness at least used to beat his wife. It doesn't say that in so many words, but a reasonable person, upon hearing the question, will conclude that the witness is or was a wife-beater. If the witness never beat his wife or was never even married, then the question in (35) does not make sense. Having a wife and beating her (at some point) are preconditions to the proper use of (35) as a question.

Finally, as the example in (35) illustrates, accommodation might be used strategically to communicate information without asserting it. We would expect, then, that a study of accommodation would reveal something about backchannel communication and its relation to discourse structure.

A sentence presupposes something, presumably a proposition, when the sentence can be true or false only when the proposition is true. The idea has been around at least since Frege's work on language in the nineteenth century. The philosopher P. F. Strawson was a particularly clear-spoken defender of presuppositions. He was vexed by Bertrand Russell's analysis of sentences like

(36) The king of France is bald.

Russell's analysis said that an utterance of the sentence in (36) was true when the following three conditions hold:

- There is at least one king of France.
- There is not more than one king of France.
- There is nothing that is a king of France and is not bald.

Since there is no king of France, then any contemporary utterance of (36) is false, according to Russell. An Englishman living in, say, 1710 might be able to utter the sentence in (36) truthfully, depending on whether Louis XIV was bald.[1]

Russell's analysis annoyed Strawson, who protested,

> Now suppose someone were to say to you with a perfectly serious air: "The king of France is wise." Would you say, "That's untrue"? I think it is quite certain that you would not. But suppose he went on to ask you whether you thought that what he had just said was true, or was false; whether you agreed or disagreed with what he had just said. I think you would be inclined, with some hesitation, to say that you did not do either; that the question of whether his statement was true or false simply *did not arise*, because there was no such person as the king of France. You might, if he were obviously serious (had a dazed astray-in-the-centuries look), say something like: "I'm afraid you must be under a misapprehension. France is not a monarchy. There is no king of France." And this brings out the point that if a man seriously uttered the sentence, his uttering it would in some sense be *evidence* that he *believed* that there was a king of France. (Strawson 1950)

According to Strawson, then, example (36) fails to be either true or false. The following sentence,

(37) The king of France is not bald.

would also fail to have a truth value, since the *existence presupposition* that use of *the king of France* carries with it would be unfulfilled.

At first glance, Strawson is right about the existence presupposition of *the king of France is bald*. Both the sentence and its negation in (37) appear to depend on the truth of

(38) The king of France exists.

This is one of the defining characteristics of presupposition.

According to Russell, the sentence in (37) as ambiguous between

(39) a. It is not the case that the king of France is bald.
    b. The king of France is unbald.

Russell would say that the interpretation in (39a) is true; it is false that the king of France is bald. The interpretation in (39b) is false for the same reason that (36) is false: there fails to be at least one king of France.

I suspect that Russell was thinking about sentences like

(40) The largest prime number is eleven.

The sentence in (40) is false, according to Russell. According to Strawson, however, the sentence in (40) is neither true nor false because the largest prime number doesn't exist.

Let's now turn to accommodation. Suppose that I say to you,

(41) My uncle lives in New Mexico.

This sentence carries an existence presupposition, namely, that I have an uncle. Chances are, you don't know me, so you probably didn't know that I have an uncle. Many people don't have uncles. But by virtue of my using the sentence in (41), you now know that I do have an uncle and that he lives in New Mexico. Your information state changes so that you *accommodate* the presuppositions of my utterance.

Now, the presuppositions of an utterance are usually taken to be part of our common knowledge. For example, you know that in order for me to have an uncle who lives in New Mexico, I have to have an uncle in the first place. Equally, in order for someone to have stopped beating his wife he has to both have a wife and have beaten her.

Notice that the following sentences are quite peculiar:

(42) a. John stopped beating his wife, although he's never been married.
 b. The king of France, who happens not to exist, is bald.
 c. My uncle lives in New Mexico and is, in fact, a figment of my imagination.

The sentences in (42) are bizarre precisely because they run contrary to the normal workings of the world. On hearing one of them, one is left with the feeling that there's something to be explained. For example, what kind of social arrangement outside of marriage (common-law included) could lead to someone's having a wife?

Of course, these days we've multiplied the ways in which someone could have a wife. I think that some people suppose that a civil union between two women is not marriage but does result in the creation of wives. In this case,

(43) Joan stopped beating her wife, although she's never been married.

could be a sensible thing to say; the presuppositions would have to be that Joan is in a civil union with another woman and that civil unions aren't really marriages. We might still maintain that the sentence in (42a) presupposes that John is married (and not in a civil union) on the grounds that heterosexual couples can only marry, whereas homosexual couples can only join in civil unions.

This highlights how presuppositions ultimately depend on how the world works. The example in (42a) and its contrast with (43) would have been incoherent a generation ago, simply because civil unions didn't exist. Social institutions have changed over the years, resulting in a supposed distinction between marriage and civil union. I suspect that in less than another generation, the preceding paragraph on marriage and civil unions will seem quaint.

Now, let's return to my simple utterance that my uncle lives in New Mexico. The peculiar thing is that my uncle was not part of our common knowledge until I uttered the sentence, so the fact that I have an uncle could not possibly have been part of common knowledge.

When common knowledge is not coordinated, we often have misunderstandings that need to be repaired. The conversation goes briefly off-track, and something must be done to coordinate. With accommodation, however, this repair is virtually automatic; in fact, it is as though the presupposition of the utterance had simply been asserted.

The misunderstandings considered in the previous section arose when the interlocutors were playing different games. That is, their understandings of the situation were such that they constructed different game trees and found themselves unable to reconcile the game they thought they were playing with the way things turned out; hence, the confusion on the part of the players. Misunderstandings, then, are a kind of coordination problem in the form of game trees.

Presuppositions are different. A presupposition is a precondition for playing the game at all. Consider playing a game like bridge. In order to play bridge, a certain number of preconditions must be satisfied, for instance, there have to be four players, and the deck of cards must be complete. If any of these preconditions fails, then the resulting game isn't bridge; it might be an approximation of bridge, but it isn't genuine bridge.

Similarly, a language game might place certain preconditions on the model or the players before play can begin.

Let's return to the verification games between Abélard and Eloïse (see chapter 5). The players are given the sentence in (41) and a model, M:

G(my uncle lives in New Mexico; M).

Of course, certain choices are fixed beforehand. The adjective *my* has to refer to the speaker, but basically Eloïse has a winning strategy if and only if she can produce an individual who is my uncle and who lives in New Mexico. Otherwise, according to the game conventions, Abélard wins and the sentence is false. Suppose Eloïse looks in the model and finds the right fellow, a portly gentleman who lives in New Mexico and is my uncle. She wins. Suppose she looks in the model and finds that I have three uncles, two of whom live in Omaha and one in Cleveland. She can produce an uncle, but not one that lives in New Mexico. She loses.

But now suppose she looks in the model and can't find a person that counts as my uncle. She has no hope of winning, but should Abélard be allowed to win the game by default? Eloïse would complain that she was at an unfair disadvantage, that the game only makes sense if I have an uncle. So let's stipulate that before the game can begin, Eloïse needs to have a winning strategy on

(44) I, Robin Clark, have at least one uncle.

which can be indicated by

G(my uncle lives in New Mexico; M)$_{G(\text{I have an uncle};\ M)^*}$,

where the subscript G(I have an uncle; M)* indicates that Eloïse has a winning strategy on *I have an uncle* in model M.

Now, when does this happen? If I say, "I have an uncle," it doesn't presuppose that I have an uncle; it asserts it. I might be lying, and the sentence is false. So not all occurrences of a noun phrase presuppose the existence of something that witnesses the noun phrase.

Here, we return to the idea that we are satisficers. As with H. Clark's analysis of the mutual knowledge paradox, assume that linguistic agents use heuristics to approximate rational behavior. One potential heuristic is

(45) **Topic Heuristic**
     If the speaker is talking about some object or objects, X, then take the existence of X as presupposed.

The heuristic in (45) is really an example of cooperation. The hearer assumes that the speaker is being truthful; if the speaker is talking about something, it's both sensible and polite to assume that something exists. Of course, the principle in (45) is only a heuristic, not a hard-and-fast constraint. We can certainly disagree about what exists, but doing so incessantly will lead to a breakdown in the conversation. Equally, the speaker might be spinning stories, and the hearer might play along; in other words, both participants would behave as though certain things exist, even though they know full well that those things don't exist.

This brings up an important idea that I want to emphasize. The topic heuristic in (45) is most emphatically *not* intended as some kind of rule of grammar. Instead, it is a characterization of something *outside* the grammar itself. I intend it to be derivable from the rational use of grammar as a tool. That is, (45) should be derivable from the theory of how grammar is used rationally to signal meanings. The game-theoretic analysis of language use should derive (45) from first principles concerning the array of choices facing the speaker and the hearer as well as their interest in maximizing their utilities.

Consider the following:

(46) Bill stopped his subscription to the *Inquirer*.

Anyone who grasps the meaning of the verb *stop* knows that it requires two states of the world, arranged in an ordered sequence. For example (46), there was a state of the world where Bill subscribed to the *Inquirer*, followed by a state of the world where Bill does not subscribe to the *Inquirer*. This is simply what it means to stop doing something. A cooperative listener will therefore adopt as true the proposition that Bill used to subscribe to the *Inquirer*. There is no need to stipulate a special grammatical rule linked to the verb *stop* that updates the discourse model with this new fact.

Let's return to the question of the existence of my uncle from New Mexico. Example (41) is more about me than about my uncle. If I have no uncle, the sentence in (41) is not about anything and *I have an uncle* is false. If *I have an uncle* is false, then the presupposition to *my uncle lives in New Mexico* fails. In that case, the game is called off and returns a third truth value:

G(my uncle lives in New Mexico; M) = **suspended**.

No one wins or loses; the game simply isn't played. So **suspended** winds up being a third truth value; it holds when a sentence is neither true nor false.

Now what about accommodation? In a normal conversation, the utterances are sincere, that is, the speaker intends to assert them and intends them to be true. Thus, we have the following verification game:

(47) G(my uncle lives in New Mexico; M)$_{G(\text{I have an uncle}; M)^*}$.

The game in (47) is a verification game, a game between Eloïse and Abélard. If the sentence is uttered sincerely, Eloïse has a winning strategy on the main game:

(48) G(my uncle lives in New Mexico; M).

But according to the topic heuristic in (45), the discourse model, M_D, must be updated with the results of the game:

(49) G(I have an uncle; M).

Recall that M_D is a data structure maintained during verification games (see rule (**R.some/a**) in chapter 5). Since cooperation requires us to assume that Eloïse wins the game in (49), we might simply stipulate some entity to fill in for my uncle from New Mexico. I could then add,

(50) He used to work in an observatory.

and the pronoun could be taken to be my uncle from New Mexico, since an appropriate entity, eligible to fill in for the pronoun, exists in the discourse model.

The process of accommodation is quite widespread; it is a common way for interlocutors to coordinate their common knowledge. Consider, for example, the following text:

(51) Last year when I was in Paris, I took a cab from Le Marais to the Place Charles de Gaulle. The driver was Senegalese.

Although the cab driver is new to the discourse, I'm able to refer to him using a definite description, which usually requires that an element already be in the discourse. If I replace *the driver* by *a driver*, which is the usual way of introducing a new element to the discourse, the text becomes a bit peculiar:

(52) Last year when I was in Paris, I took a cab from Le Marais to the Place Charles de Gaulle. A driver was Senegalese.

Whoever "a driver" is, it isn't the driver of the cab I took, unless it had multiple drivers who took turns driving, which seems unlikely. What happens in this kind of case? Since cabs normally have drivers, my introduction of the cab in the first sentence allowed for the accommodation that it had a driver. Thus, the hearer cooperatively parked the cab driver in the discourse model along with the cab.

In summary, participants in a conversation require a store of common knowledge if communication is to succeed. But common knowledge is, in fact, simultaneously unachievable and indispensable. The solution is to assume that linguistic agents are capable of bounded rationality. They use heuristics to approximate common knowledge.

Given that common knowledge can only be approximated, we would expect that communication should fail in some instances, namely, misunderstandings. In these cases, one of the participants in a conversation cannot work out a coherent game tree, and the conversation must be repaired. Misunderstandings may be quite common but unnoticed.

The failure of common knowledge can be repaired by accommodation. Sometimes utterances make presuppositions that are not part of common knowledge. In these cases, the interlocutors can work out a game tree; they cooperatively update their discourse models with the presupposed information.

Reconciling the Assumptions

I can now state my fundamental hypothesis, a variation of Putnam's Hypothesis of the Universality of the Division of Linguistic Labor (see chapter 3):

> The meaning of any linguistic expression is a function of its use in context. In particular, speakers and hearers use grammar to signal meaning, and this use is based on principles of bounded rationality. Use is *publicly available* and regulated by conventions. Although mental representations play a causal role in use, meaning is ultimately socially regulated.

I have also argued that knowledge can never be shared with anyone else. It might seem as though this fact should be fundamentally at odds with my hypothesis. The position I've explored in this chapter seems to skirt full solipsism. Am I claiming that ultimately we are brains in SUVs, alone with our mental representations?

The short answer is, of course, no. Simply because we are not mind readers does not imply that we are fundamentally cut off from each other. Our public linguistic behavior allows us to coordinate around focal points (see chapter 9) and thus to regiment our model of the world and our models of each other. In fact, without this public coordination, language and perhaps thought itself would be ultimately contentless. Language and thought without content would simply not exist.

Further Reading

The idea of bounded rationality is due to Herb Simon. Most important are his 1955 and 1956 papers, both collected in Simon (1982). He produced many other papers on the topic. Simon (1982) gives a representative sample. Rubinstein (1998) gives some formal models of bounded rationality. Bounded rationality is one of the central topics of behavioral game theory; see Camerer (2003).

The coordinated attack problem is from Fagin et al. (1995). As usual, I have added my own flourishes. The coordinated attack problem is quite general; computer networks, for example, must find some reliable method of overcoming the problem. As noted, there is no perfect solution; common knowledge is unattainable.

Alice and Buddy's problem with definite descriptions is adapted from H. Clark (1992). In that book and in H. Clark (1996), Clark made important contributions that I believe, have not received enough attention among linguists.

My thinking about definite descriptions has also been influenced by Neale's (1990) defense of Russell's account of descriptions. I think that Neale is largely correct in his Gricean account of reference. I return to this point in chapter 9, when I take up focal points in earnest. Another influence on my thinking about both definite descriptions and pronouns is Evans (1982; 1985). Definite descriptions provide a vast laboratory for game theory.

Labov's work on misunderstandings is discussed in *Principles of Linguistic Change* (2010). I am grateful for the advance look before publication. I think that strategic decision making plays a role even at the level of phoneme perception as well as during speech production. We make thousands of low-level decisions about articulation during both production and perception. This should be a rich area for decision-theoretic experimentation.

Presupposition has received a vast amount of attention over the years, much of it centered on the problem of presupposition projection, whether a main clause inherits the presuppositions of an embedded clause. A good discussion of work on presupposition can be found in Beaver (1997). My brief discussion of the related question of accommodation relies on Lewis (1983). A great deal of work has been done on accommodation over the past two decades, and a number of important principles have been discovered. See Beaver and Zeevat (2007) for an overview.

7 Lexical Games

The previous chapter discussed some communication games and how speakers and hearers coordinate on descriptions like

(1) the dance troupe performing at the Prance this weekend.

The question that Alice and Buddy had to address is whether they could coordinate on the referent of the description in (1) when the context provided two possible referents for the description.

This situation seems common enough; almost everything we say or write is ambiguous, yet we're usually unaware of the ambiguity. It's easy to say that the context somehow decides the ambiguity. We have the notion that we somehow use information from the context to decide what was meant by an ambiguous sentence or word. If I use a word like *pen*, for example, you usually know whether I mean a writing instrument that uses ink or an enclosure for animals.[1] Given that ambiguity is ubiquitous in day-to-day speech, how are we are able to coordinate our linguistic behavior?

The extensive game illustrating Alice and Buddy's coordination problem is repeated in figure 7.1. Note that the game in the figure is not an accurate model of the communication problem; this is corrected later.

Games for Finding Words

As noted, *pen* has quite a few different interpretations, but for the moment, suppose that *pen* is just two-ways ambiguous, with the following interpretations:

(2) a. an instrument for writing or drawing with ink,
    b. an enclosure for animals.

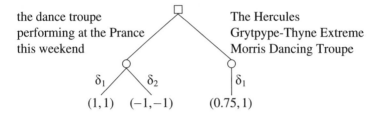

□ = Alice
○ = Buddy
δ_1 = The Hercules Grytpype-Thyne Extreme Morris Dancing Troupe
δ_2 = The Hans-Erni Spruengli Troupe of Appenzell Sword Dancers

Figure 7.1
Simple (Misleading) Communication Game

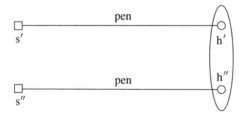

Figure 7.2
Speech Event

Now, suppose someone says the word *pen* out of context. Perhaps you overhear the single utterance "pen!" with no further information. All else being equal, you can't tell whether the speaker intended (2a) or (2b).

Figure 7.2 shows the problem. Suppose the speaker can be either in state s', the state of intending to use the word *pen* to refer to a writing instrument, or in state s'', the state of intending to use *pen* to refer to an enclosure for animals.

Of course, uttering "pen" has had an effect on the hearer. Let h' indicate that the hearer has understood it to mean a writing instrument, and let h'' indicate that the hearer has understood it to indicate an enclosure for animals. The hearer knows that by uttering "pen" the speaker intended to indicate one or the other. But the information at hand does not define which kind of thing the speaker intended. Thus, the speaker's

Lexical Games 217

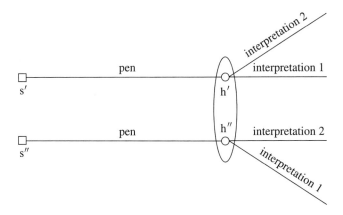

Figure 7.3
Speech Event and Hearer's Choice

utterance "pen" has placed the hearer in an information set consisting of the states h' and h'', which are encircled in the figure to indicate that they form an information set.

The situation in the figure suggests that the hearer faces a decision if she wants to resolve the ambiguity implicit in the information set $\{h', h''\}$. Suppose the speaker intended by uttering "pen" something involving a writing instrument; call this interpretation 1. For present purposes, interpretation 1 could be almost anything involving a writing instrument. It could be a particular pen, the concept denoted by *pen* in its writing instrument guise, or even the command "Bring me something to write with!" For the moment, we can be totally agnostic about what the content of interpretation 1 is; we need only assume that it exists and is available to the hearer.

Equally, the speaker could have intended by uttering "pen" something about enclosures for animals. We can again remain indifferent as to what exactly the speaker wanted to say. Let's call this interpretation 2 and remain vague as to whether the speaker intended to indicate the concept, a particular enclosure, or perhaps a command of some sort, for example, "Put the animals in that pen!"

Figure 7.3 shows a modified diagram of the speech event in figure 7.2 to reflect the hearer's decision. Recall that s' indicates that the speaker intended to signal something about a writing instrument. If the hearer knew this, then she would know that she was in information state h', the one induced in the hearer by recognizing the speaker's intention to refer to writing instruments. In this case, she would correctly pick interpretation

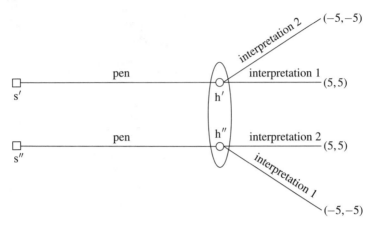

Figure 7.4
Speech Event, Hearer's Choice, and Utilities

1; choosing interpretation 2 would be a mistake, since the speaker did not intend her to glean that interpretation.

If the speaker was in state s'', he intended to signal something about animal enclosures. If the hearer knows this, then she is definitively in information state h'', the state of recognizing the speaker's intention to talk about animal enclosures. In this case, she should rationally choose interpretation 2 rather than interpretation 1. This observation allows us to flesh out the decision problem implicit in figure 7.3. Figure 7.4 adds utilities to the hearer's choice problem.

In assigning utilities to the speaker and hearer, I assume that they share an interest in successful communication. The speaker wants the hearer to recognize what he intends to signal, and the hearer wants to recognize the speaker's intention. Thus, in figure 7.4, the interests of the speaker and hearer coincide. This doesn't have to be the case. For example, the speaker might have an interest in getting the hearer to make the incorrect choice (when he lies, for example) or to fail to fully interpret his intention (dissimulation). Equally, the speaker might prefer to say as little as possible, to use the minimal means to signal the most information, while the hearer might want the speaker to say more and thus make his intention clearer. Game theory provides sufficient tools to build models of all these kinds of behavior.

Assume that the speaker and the hearer want to coordinate their linguistic behavior. (See the discussion of coordination games in chapter 4.) Also assume that the speaker and the hearer prefer to coordinate on the

		Speaker	
		Interpretation 1	Interpretation 2
Hearer	Interpretation 1	5, 5	−5, −5
	Interpretation 2	−5, −5	5, 5

Figure 7.5
Communication Game in Strategic Form

speaker's intended interpretation. Indicating the speaker's intention and the hearer's choice as a pair, we get the following ranking:

(interpretation 1, interpretation 1) > (interpretation 1, interpretation 2),

(interpretation 2, interpretation 2) > (interpretation 2, interpretation 1).

The speaker and hearer both get 5 points when the hearer chooses the interpretation that correctly corresponds to the speaker's intention. They both get a penalty when the hearer chooses incorrectly.

You might suppose that this coordination problem could be shown as a game in strategic form (figure 7.5). But a little reflection should convince you that the form shown in figure 7.5 won't work because it does not accurately reflect the speaker's decision problem. The problem is that in producing a form like *pen*, the speaker also faces a strategic decision problem; the speaker must decide whether to use the simple form *pen* to convey his meaning or try to formulate another expression—"something to write with," for example—that will make his intended meaning clear to the hearer. Notice that the speaker could actually be faced with a further game, the problem of generating a new form for an interpretation, given a local context. For convenience, this new game is represented as

G(interpretation; Speaker; Hearer; form; Context),

where the game is understood as being the particular one of the speaker's choosing an alternative to some form that carries the intended interpretation to a particular hearer (or hearers), given a context.

The game in figure 7.6 is a *game of partial information*. The game tree is depicted differently than earlier ones in that it grows sideways. From now on, I follow the convention that game trees grow from left to right, which invokes something of the temporal dimension of language.

Recall from chapter 4 that in a game of perfect information, each player can immediately observe her opponent's choices. This would be a game like chess or checkers, where each player's choice is announced in

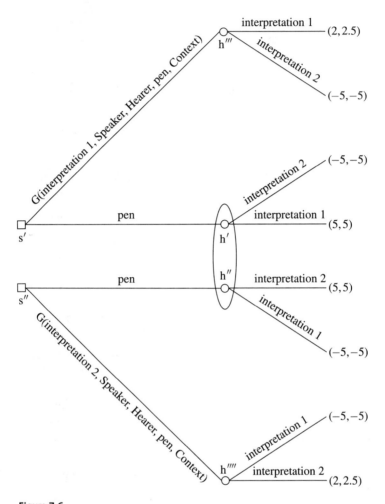

Figure 7.6
Communication Game of Partial Information

turn, and the board is changed to reflect that choice. (First-order logic can be thought of as a game of perfect information; see chapter 5.) In a game of incomplete information, the players do not have information about their opponents' choices. An example of this kind of game is the Holmes-Moriarty game (see chapter 4). In a game of partial information, a player can announce his choice, but that announcement does not completely inform his opponent's choice. Thus, when the speaker uses the word *pen*, the hearer is left to sort out exactly what he intended: writing instrument or animal enclosure?

Lexical Games

Now, some things have been added in figure 7.6. First, the speaker must make a choice between using the word *pen* and sending the hearer to the information set $\{h', h''\}$, or playing a further game to formulate some other way of linguistically encoding his intended meaning. The exact game the speaker will play is contingent on what his intentions are; if he intends to denote a writing instrument—that is, if he is in information state s'—then he will play

G(interpretation 1; Speaker; Hearer; *pen*; Context).

He will play a game to choose a form other than *pen* to encode his intended meaning of 'something to write with'. If the speaker intends to denote an enclosure for animals—if he is in information state s''—then he will play

G(interpretation 2; Speaker; Hearer; *pen*; Context),

where he must choose an alternative encoding for *pen* in order to indicate 'animal enclosure'.

Suppose that the speaker chooses to express the intended interpretation 'writing instrument' using an alternative to *pen*. He plays the new game and selects some expression—perhaps "that writing instrument over there"—and transmits it to the hearer. His utterance then sends the hearer to information state h''', which is the state where she correctly and unambiguously recognizes the speaker's intention to refer to a writing instrument.

On the other hand, suppose that the speaker decides to formulate an alternative to *pen* in order to express the meaning 'animal enclosure'. Then he plays a game to decide what expression best suits his intended interpretation; his utterance sends the hearer to information state h'''', where she recognizes his intention to refer to an animal enclosure.

Orderly Communication and Utility

Although both expressions that are selected as alternatives to *pen* are assumed to be unambiguous in the context, I have assigned them relatively low utility. If he could get away with it, if he could get the hearer to recognize his communicative intention, then the speaker would prefer to say "pen" because it is short and there is little work involved. In fact, if I could get away with it, I would communicate using only the monosyllable "uh" because it's so easy to say. Sadly for me, I often have to say more in order to distinguish my communicative intentions; sadly for my audience, they often have to listen to me ramble on as I try to make myself understood.

Now, given that he must find an alternative expression to encode his intended meaning, the speaker must do the work of selecting an alternative form and then articulating it. Equally, the hearer would prefer the speaker to be brief and to present as short an utterance as possible, since she must process and interpret his utterance. The hearer, too, would prefer that the speaker use the word *pen* if possible.

An implicit theory of value is encoded in the utilities shown in figure 7.6:

(3) a. Communication is paramount. The speaker prefers the hearer to recognize correctly his intention in uttering a linguistic expression, and the hearer prefers to recognize correctly the speaker's intention.
   b. Both the speaker and the hearer prefer to minimize the work they must do in communicating. The speaker wishes to minimize the number of choices he must make, and the hearer wishes to minimize the amount of processing she must do in working out the speaker's intentions.

The condition in (3a) is due to the philosopher H. Paul Grice, who sought to formulate the mechanics of meaning in his paper "Meaning":

Perhaps we may sum up what is necessary for A to mean something by x as follows. A must intend to induce by x a belief in an audience, and he must also intend his utterance to be recognized as so intended. But these intentions are not independent; the recognition is intended by A to play its part in inducing the belief, and if it does not do so something will have gone wrong with the fulfillment of A's intentions. (Grice 1957, 383)

Grice's formulation clearly involves strategic thinking. In the same paper, he wrote, "'A meant something by x' is roughly equivalent to 'A uttered x with the intention of inducing a belief by means of the recognition of this intention'." When the speaker plays a game to decide how to encode his desired meaning, he is working out how best to get the hearer to recognize his intention. Equally, the hearer, in processing the utterance, must play a similar game to work out what the speaker must have intended by saying something in a particular way. The conditions in (3a) and (3b) follow from some general principles about how interlocutors organize their conversation.

Notice that the desire for accurate communication can be undermined in a variety of ways. The speaker may be lying or dissimulating. The speech might be ceremonial and thus not communicative in the sense intended in (3a). But these are surely different language games from the game we are considering here, where the speaker intends to be understood in a fairly direct fashion.

The condition in (3b) is a kind of least-effort condition that says both the speaker and the hearer normally intend to minimize the amount of work they have to do in producing and processing speech. Grice (1975) gives a number of principles for orderly communication. First there is an overarching principle, called the cooperative principle:

(4) *Cooperative Principle*
Make your conversational contribution such as is required at the stage at which it occurs, by the accepted purpose or direction of the talk exchange in which you are engaged.

It is assumed that the participants in a conversation are always trying to obey the cooperative principle (which is somewhat vague). The principle is supported by a small set of maxims, shown in figure 7.7. The maxims themselves seem like good advice to rational agents on how to structure their conversational practice. Grice intended them as such; his contention was, in brief, that pragmatic meaning arises from the *rational use* of language. The maxims are not part of the grammar of any language but rather describe how any rational agent would *use* a grammar to signal meaning.

It is sometimes impossible to simultaneously obey all the conversational maxims and the cooperative principle. When a speaker violates one or more of these, Grice argued, it results in some additional meaning being signaled. This additional meaning is called an *implicature*.

The overarching architectural element is the cooperative principle in (4). In violating one of the maxims, a speaker implies that although she really wants to be cooperative, she had to violate the principle because obeying it would be conversationally unacceptable. The hearer, then, is left to work out why the speaker had to violate the principle.

Let's take one of Grice's examples and show how it is related to the condition in (3b) that states that the speaker and hearer jointly prefer to minimize their work. For instance, the manner maxims regulate the way in which information is stated. We are concerned with manner because it correlates with the effort that must be expended in speaking and processing. The manner maxims imply a principle that minimizes the work done by both the speaker and the hearer, particularly when combined with the quantity maxims (see figure 7.7). The latter advises speakers to say as

> Maxims supporting the cooperative principle (Grice 1975):
>
> **Quantity**
>
> 1. Make your contribution as informative as is required (for the current purposes of the exchange).
> 2. Do not make your contribution more informative than is required.
>
> **Quality**
>
> Try to make your contribution one that is true.
>
> 1. Do not say what you believe to be false.
> 2. Do not say that for which you lack adequate evidence.
>
> **Relation**
>
> Be relevant.
>
> **Manner**
>
> Be perspicuous.
>
> 1. Avoid obscurity of expression.
> 2. Avoid ambiguity.
> 3. Be brief (avoid unnecessary prolixity).
> 4. Be orderly.

Figure 7.7
Grice's Conversational Maxims

much as they need to say and no more, so that it's easy to produce and easy to process.

Grice invites us to compare the following:

(5) a. Miss X sang "Home Sweet Home."
    b. Miss X produced a series of sounds that corresponded closely with the score of "Home Sweet Home."

Example (5a) is a fairly minimal way of saying someone sang "Home Sweet Home," whereas (5b) is long and complicated, says more than the hearer needs or wants to know, and says it in an odd way. Both sentences in (5) have the truth condition that some particular person sang a particular song, but (5b) carries with it the implication that Miss X produced a defective performance. One can't help but think that the only reason someone would produce a convoluted utterance like (5b) is that they don't want to say outright, "Miss X's performance of "Home Sweet Home" stank!" Indeed, by saying things in such a prolix way, the speaker invites the hearer to come to this conclusion, while still being able to deny that he meant that.

The speaker could explicitly deny that he intended this meaning, a process called cancellation, which is one thing that distinguishes implicatures from entailments (see chapter 1, definition (10)):

(6) Miss X produced a series of sounds that corresponded closely with the score of "Home Sweet Home." It was, in fact, the most delightful and entrancing performance of the piece that I have ever had the pleasure of hearing.[2]

The bloated utterance in (6) is not a contradiction, although the first sentence implicates (but does not entail) that Miss X's performance was not very good. Compare it with

(7) Precipitation in the form of water droplets is falling at the moment, but it's not raining.

This statement is decidedly bizarre, since rain is exactly precipitation in the form of water droplets.

The principles in (3a) and (3b), then, are not absolute, but they do serve to define a class of language games that corresponds to a garden-variety use of language as a simple means of communication. As Grice's example shows, when we violate these principles, we do so for principled reasons that invite our audience to speculate and come to conclusions about the reasons for the violation. Grice's maxims fit into the game-theoretic framework; we can encode their effects using utilities. In the normal case, such as deciding whether to use *pen* or *writing instrument*, Grice's maxims point toward assigning higher utility to *pen* than to *writing instrument*. The context provides probabilities, and game theory describes how a rational agent would solve the game. Implicature is considered in more detail in chapter 8.

I would argue that the assignment of utilities according to the principles in (3a) and (3b) is a fair reflection of ordinary preferences. Notice that the result amounts to a Gricean theory of lexical access. It's possible that in some situations I would assign utility differently; if I'm trying to mislead you or to cover up something, then I won't want you to recognize my intention to mislead. But this just shows that in usual discourse there are a variety of games that could be played. For now, let's focus on simple games of asserting and denoting.

Playing the Odds

There's one more element that should be added to figure 7.6 to make it into a proper game. Notice that the diagram consists of two game trees. One game tree is rooted in s', which represents the speaker's intention to refer to a writing instrument, while the other game tree is rooted in s'', which represents the speaker's intention to refer to an enclosure for animals. It is useful to think of the speaker as being in one state or another with a certain probability.

Recall that in chapter 5, I introduced a player, nature or chance, to work out certain ambiguities. Nature determined whether Eloïse or Abélard got to move at a certain point, and that would determine whether a sentence was interpreted one way or another. A similar device is useful here. In figure 7.8, I've suppressed the trivially bad choices that the hearer could (but shouldn't) make at nodes h''' and h'''', since they add nothing to the discussion. Chance determines whether the speaker is in state s' or state s''. That is, with probability p, chance places the speaker into s', and with probability p' the speaker is in s''. For present purposes, assume that p and p' sum to 1, so that

$$p' = 1 - p.$$

That is, in the situation at hand, the speaker will either refer to a writing instrument or to an enclosure for animals; the probability of the two events should sum to 1.

What happens if p is 0.5? The probability of the speaker's intending to mean 'something to write with' is the same as the probability of his intending 'animal enclosure', since, by the above equation, p' must also be 0.5. Recall that both the speaker and the hearer want to maximize their expected utility, where expected utility is the product of the utility of an outcome and its probability:

$$EU(g(a)) = p_{g(a)} \times U(g(a)),$$

where $g(a)$ is the outcome associated with an action a, $U(g(a))$ is its utility, and $p_{g(a)}$ is the probability of that outcome (see the discussion concerning figure 4.12).

The expected utilities for the speaker and the hearer in this game are shown in table 7.1. G, represents

Lexical Games 227

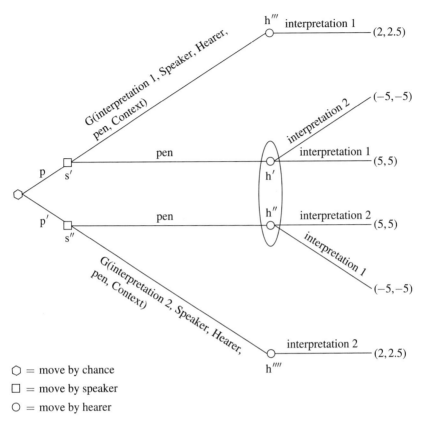

○ = move by chance
□ = move by speaker
○ = move by hearer

Figure 7.8
Lexical Choice as a Game of Partial Information

Table 7.1
Expected Utilities when $p = p'$

| | Action | | Utility | | Prob- | Expected Utility | |
State	Speaker	Hearer	Speaker	Hearer	ability	Speaker	Hearer
s', h'''	G_1	Int 1	2	2.5	0.5	1	1.5
s', h'	pen	Int 2	−5	−5	0.5	−2.5	−2.5
s', h'	pen	Int 1	5	5	0.5	2.5	2.5
s'', h''	pen	Int 2	5	5	0.5	2.5	2.5
s'', h'	pen	Int 1	−5	−5	0.5	−2.5	−2.5
s'', h''''	G_2	Int 2	2	2.5	0.5	1	1.5

G(interpretation 1; Speaker; Hearer; *pen*; Context),

and G_2 represents

G(interpretation 2; Speaker; Hearer; *pen*; Context).

The top and bottom lines of table 7.1 are the cases where the speaker chooses an unambiguous alternative to *pen*—the descriptions *something to write with* or *animal enclosure*—and the hearer is able to select the intended interpretation. The other lines in the table correspond to the case where the speaker utters "pen" and thus sends the hearer to the information set $\{h', h''\}$. Notice that in this case the hearer has no reason to suppose that interpretation 1 is more likely than interpretation 2, or vice versa; the hearer can only assume they are equally likely.

Now, to work out the expected utilities of the choices in the information set, we would multiply the utilities connected to h' by p and the utilities connected to h'' by p'. In this case, p and p' are both 0.5. Obviously, if the hearer knew she was in h', she should pick interpretation 1, and if she knew she was in h'', she should pick interpretation 2. But she can only guess which action to choose; she can anticipate being wrong half the time. So her expected utility as well as the speaker's is 0 in this case; half the time they each get 2.5, and half the time they each get -2.5.

Now, a rational agent prefers 1 to 0, so the speaker should avoid using *pen* in this case. The best strategy is as follows:

$\{(s', G_1), (s'', G_2), (h''', \textit{writing instrument}), (h'''', \textit{animal enclosure})\}.$

That is, if the speaker wants to refer to a writing instrument, avoid *pen* but formulate some alternative description. If the speaker wants to refer to an enclosure for animals, avoid *pen* and formulate another alternative description. The advice to the hearer is trivial. There should really be a choice for the hearer if the speaker messes up and uses *pen*; in this case the hearer could follow a mixed strategy, choosing interpretation 1 half the time and interpretation 2 the other half.

All of this sounds like good advice in situations that are truly ambiguous. If I'm the speaker and, in my judgment, either interpretation is just as likely as the other, I probably would try to formulate some less ambiguous description to get my meaning across. Of course, this sort of profound ambiguity—where the situation gives me no preference as to the intended reading—is hard to find. Chapter 6 mentioned the example

(8) Colin Powell endorsed Barack Obama because he's black.

as a case of genuine indeterminacy. In this case, taking the pronoun *he* as denoting either Colin Powell or Barack Obama is possible; either interpretation results in a coherent reading of the sentence, given the context. French provides a lovely example:

(9) a. La belle ferme la voile.
    The beauty closes the veil
    'The beauty closes the veil.'
   b. La belle ferme l 'a voile.
    The beautiful farm it has hidden.
    'The beautiful farm hid it.'

The two sentences *sound* exactly alike (although they're written differently), so the example is perfectly ambiguous when spoken, if clues from the context are ignored.

How often does it happen that an ambiguity is completely indeterminate? It's hard to say. Perhaps linguists notice ambiguities more than others do, even when it's clear which interpretation is intended. This habit of noticing ambiguity makes talking to linguists something of an annoyance.

Most often, there are cues in the context that will allow the hearer to work out which interpretation is intended.

Clues from the Context

Suppose you're at a meeting of the Modern Language Association (MLA). People attending a meeting of the MLA generally need things to write with but, for the most part, don't have much interest in enclosures for animals. If you hear the word *pen* in that context, you're likely to have a bias for the writing instrument interpretation.

The move by nature (see figure 7.8) should reflect this shift in judgment; technically, the *subjective probability* that is associated with the state s' should be adjusted. Subjective probability is the likelihood that someone assigns to an event like the proper interpretation of *pen*, given that person's general knowledge of the world. Since you know that people attending an MLA meeting are interested in language and writing, you can safely assume that by *pen* they mean a writing instrument and probably don't mean an animal enclosure. You could be wrong, but the odds are in your favor.

So, in figure 7.8, let's adjust p (the probability that the speaker is in state s' where he wants to refer to a writing instrument) from 0.5 to 0.7. Since $p' = 1 - p$, this means that p' (the probability that the speaker wants to refer to an animal enclosure) changes from 0.5 to 0.3.

Since the probabilities have changed, so have the expected utilities associated with the outcomes of the game. Suddenly, things look rather

Table 7.2
Expected Utilities when $p > p'$

State	Action		Utility		Prob-	Expected Utility	
	Speaker	Hearer	Speaker	Hearer	ability	Speaker	Hearer
s', h'''	G_1	Int 1	2	2.5	0.7	1.4	1.75
s', h'	pen	Int 2	−5	−5	0.7	−3.5	−3.5
s', h'	pen	Int 1	5	5	0.7	3.5	3.5
s'', h''	pen	Int 2	5	5	0.3	1.5	1.5
s'', h'	pen	Int 1	−5	−5	0.3	−1.5	−1.5
s'', h''''	G_2	Int 2	2	2.5	0.3	0.6	0.75

different from the equiprobable case. According to the expected utilities listed in table 7.2 for the information set $\{h', h''\}$, a clear winner emerges. If the speaker uses *pen* in this case, then the hearer should choose interpretation 1—writing instrument—because the expected utility for that choice is higher than the expected utility for the other choice. Seventy percent of the time the hearer (and the speaker) will get 5, but 30 percent of the time she'll get −5, so she can on average expect to get $(0.7 \times 5) + (0.3 \times -5) = 2$ for choosing interpretation 1. Now the strategy profile for the players becomes

$\{(s', pen), (s'', G_2), (\{h', h''\}, writing\ instrument), (h'''', animal\ enclosure)\}.$

That is, if the speaker wants to signal a writing instrument, he should use the word *pen*. If he wants to talk about animal enclosures, then he would be well-advised to choose an expression other than *pen*. If the hearer hears *pen*, then she should choose the writing instrument interpretation.

This result, where one option "pops out" because of a change in probability, accords well with intuition. If I'm in a particular situation like an MLA meeting, I expect to encounter certain things like writing implements, books, book bags, and so on, while I would be surprised to encounter certain other things like plows and pig pens. When I encounter a word like *pen* in such a situation, it makes sense for me to opt for the interpretation that best accords with my circumstances.

Let's change the situation slightly. Suppose that instead of attending a meeting of the Modern Language Association, we were at a congress of the Brotherhood of Hog Farmers of Superior Bottom, North Dakota. Clearly, I have manipulated the example to lower the probability that the attendees would be interested in writing instruments and to raise the

Table 7.3
Expected Utilities when $p < p'$

State	Action		Utility		Prob-	Expected Utility	
	Speaker	Hearer	Speaker	Hearer	ability	Speaker	Hearer
s', h'''	G_1	Int 1	2	2.5	0.3	0.6	0.75
s', h'	pen	Int 2	−5	−5	0.3	−1.5	−1.5
s', h'	pen	Int 1	5	5	0.3	1.5	1.5
s'', h''	pen	Int 2	5	5	0.7	3.5	3.5
s'', h'	pen	Int 1	−5	−5	0.7	−3.5	−3.5
s'', h''''	G_2	Int 2	2	2.5	0.7	1.4	1.75

probability that they would be interested in animal enclosures. Now, someone asks,

(10) How many *pens* do you have?

What sense of *pen* do they mean?

At very least, we should raise the subjective probability associated with the speaker's intent to talk about animal enclosures. In figure 7.8, the probability that chance plays s'' should be greater than the probability that chance plays s'; p' should be greater than p. So let $p' = 0.7$ and $p = 0.3$.

The expected utilities are recomputed in table 7.3. Now the interpretation of *pen* as an animal enclosure has a higher expected utility than its interpretation as a writing instrument. In fact, the strategy profile becomes:

$\{(s', G_1), (s'', pen), (\{h', h''\}, animal\ enclosure), (h''', writing\ instrument)\}.$

That is, if the speaker wants to talk about a writing instrument, he should play the game G_1 to choose an alternative description—*something to write with*—but if he wants to talk about animal enclosures, he can use *pen*.

The translation of Grice's principles into a utility ranking plus some simple probability theory yields a fine model of lexical choice. The results of the game in figure 7.8 when the probabilities p and p' are set to accord with subjective judgments of likelihood in various contexts accords very well with intuition about how expressions would be used in various contexts. The basic idea is this: We compute the expected utility of various expressions; the expression with the greatest expected utility has the highest level of activation. This means that it is the one most likely to be used in a particular context. If two or more forms have the same expected utility, then the choice between them is random, a mixed strategy.

Back to Descriptions and Common Knowledge

Let's return to the Alice and Buddy example discussed in chapter 6. Recall that the speaker, Alice, needed to decide what expression to use to refer to a dance troupe performing at the Prance Theater the coming weekend. One dance troupe, the Appenzell Sword Dancers, was scheduled but had to cancel; it was replaced by the Extreme Morris Dancing Troupe. While Alice knows this, it is unclear whether her interlocutor, Buddy, knows, or knows that she knows, and so forth.

The game for this problem is shown in figure 7.9; again the trivially bad choices at h''' and h'''' have been suppressed. The first move is up to chance, which determines with some probability (to be specified) whether Alice intends to refer to one troupe or the other. That is, there's a probability, p, that she will be in state s', where she wants to talk about the Morris Dancers, and a probability p' that she will be in state s'', where she wants to talk about the Appenzellers.

If she says,

(11) The dance troupe performing at the Prance this weekend

she puts Buddy into the information set $\{h', h''\}$; that is, there is in principle some uncertainty about which dance group she intends by her utterance.

Suppose that Buddy doesn't know there has been a schedule change. He would assign probability 0 to p, since he knows nothing about the Morris Dancers. In that case, if Alice uses the description in (11), Buddy will assign it δ_2, since the choices dominated by h' have probability 0 and therefore expected utility 0—zero times anything is zero. Thus, the choices under h'' have the highest utilities; since only a sucker would pick δ_1—and Buddy doesn't even know that the Morris Dancers exist—he's certain to pick δ_2, the Appenzellers. It follows that if Alice believes that Buddy doesn't know that there has been a schedule change, she should use the description to refer to the Appenzellers and the name of the dance troupe to refer to the Morris Dancers.

Equally, suppose that Buddy knows there has been a schedule change but mistakenly believes that Alice doesn't know. He would again assign p the value 0—he thinks that Alice would never want to refer to the Morris Dancers. The outcome of the game is that again he would associate

Lexical Games

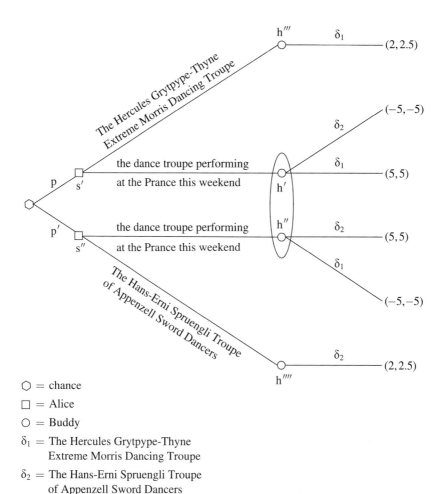

Figure 7.9
Game of Partial Information for Descriptions

the Appenzellers with the description in (11). So if Alice believes that Buddy knows about the schedule change but doesn't know that she knows, she should behave as before and use the description to refer to the Appenzellers and the name of the dance troupe to refer to the Morris Dancers.

Suppose that they both know about the schedule change and know that the other knows, and so forth. In this case, they would both assign 1 to p and 0 to p'; this case is the mirror image of the previous two cases. Alice should use the description to refer to the Morris Dancers and the name of the troupe to refer to the Appenzellers.

What about the intermediate cases? These are cases where neither p nor p' are 0, meaning that there is some chance that Alice could use the description in (11) to refer to either dance troupe. The solution here has the same basic structure as the solution for the lexical game for *pen*. If $p > p'$—if Alice is more likely to refer to the Morris Dancers, and both Alice and Buddy know (or at least believe they know) this—then she can use the description "the dance troupe performing at the Prance this weekend" to refer to the Morris Dancers, and she should use the name of the troupe to refer to the Appenzellers. If $p' > p$, then the opposite holds: Alice can use the description to refer to the Appenzellers, but she should use the full name to refer to the Morris Dancers.

Equilibrium Selection and Implicature

Now, let's assess this result. Note that for each of the games discussed so far in this chapter, there are actually multiple Nash equilibria. I've discussed one equilibrium strategy for each game, in fact, the payoff-dominant or Pareto-dominant strategy (see figure 4.23). This is just the strategy that pays off at least as well or better than the other strategies.

Notice that the description game in figure 7.9, for example, has another equilibrium strategy. Alice could always choose to use the full name for each group, and Buddy would then choose the correct referent for each name. Neither the speaker nor the hearer have any incentive to unilaterally defect from this strategy. In fact, this strategy has the advantage of having less risk associated with it. Suppose, for instance, that Alice uses the description and Buddy makes a mistake—his assessment of the probabilities is significantly different from hers—then both the speaker and the hearer risk a misunderstanding. Using the names is the risk-dominant strategy (see figure 4.24), since it avoids the possibility of getting the sucker payoff that exists when Buddy makes a (potentially erroneous) choice in the information set.

We have returned to the problem of equilibrium selection, discussed in chapter 4, combined now with bounded rationality, discussed in chapter 6. Suppose that Alice and Buddy know each other well and are prone to trust in each other's judgment about making choices. In this case, they are likely to select the Pareto-dominant equilibrium. On the other hand, if Alice and Buddy are uncertain about each other's knowledge or

Lexical Games

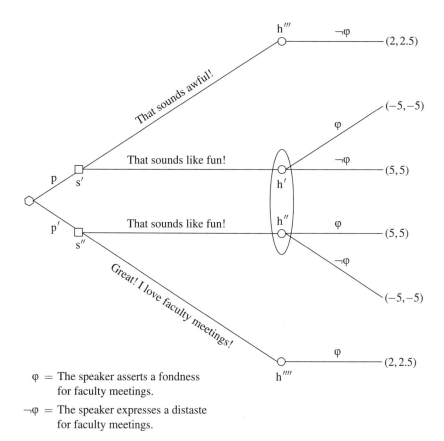

φ = The speaker asserts a fondness for faculty meetings.

¬φ = The speaker expresses a distaste for faculty meetings.

Figure 7.10
Irony Game

don't know each other particularly well, they are likely to prefer the risk-dominant equilibrium.

In general, the better the two players know each other, the more likely they are to use the Pareto-dominant strategy profile. Consider irony, which often involves the implicit negation of what was said. For example, suppose that the chair of my department calls a faculty meeting to discuss the allocation of office supplies. One of my colleagues says,

(12) That sounds like fun!

We can treat this case as a game of partial information (figure 7.10). My colleague's assertion sends me to an information set where it is unclear whether he meant the literal content of *that sounds like fun!* or something close to its opposite. He could have made his opinions clearer by saying something like

(13) That sounds awful!

if he really dislikes faculty meetings, or

(14) Great! I love faculty meetings!

if he sincerely enjoys them.

Now, I know my colleague well enough to know that he doesn't like faculty meetings. So I can safely assume that p is much greater than p' in figure 7.10. If we didn't know each other at all, then it is unlikely that he would have elected to express his meaning using irony; instead, he would have selected the risk-dominant strategy of being literal in his expression. Thus, the better we know each other, the more likely we are to do things like express ourselves nonliterally. David Sally, an economist who has worked on this problem, had the idea of adding a "sympathy" constant λ to the utility associated with the nonliteral meaning. Of course, only people who know each other well would know when to add the λ to the utility. This, by the way, explains why irony almost inevitably fails on the Internet. Posting an ironic comment to an online forum invites misunderstanding, since many of the readers of the comment lack sufficient knowledge of the writer's usual attitudes.

What makes irony a payoff-dominant strategy? Presumably, the solidarity signaled by the use of irony adds to its utility. My colleague and I are at one in our misery at the prospect of yet another faculty meeting, and we've had the pleasure of a bit of shared subversion. Never underestimate the value of community!

Another case where nonliteral, payoff-dominant meanings are preferred is in highly conventionalized exchanges. The following example was discussed in chapter 4:

(15) *Man*   Do you have a watch?
   Woman   Yes.
   (*Long pause*)
   Man (*irritated*)   Well?
   Woman   Oh! I left it at home.

The man literally asks the woman an information question, but he intends to get her to tell him the time. She fails to pick up on the nonliteral content of his utterance. Another example is

(16) Could you pass the salt?

Virtually everyone recognizes this as a request for the salt and not an inquiry into the hearer's capabilities. If you ask me whether I can pass the salt and I say yes and do nothing else, you are likely to be irritated. This sort of question is a highly conventionalized form of politeness that any speaker of the language can safely be taken to understand. That such conventionalized politeness sometimes fails is testimony to the fact that we are indeed bounded agents, prone to error and miscoordination.

Down the Garden Path

The previous section discussed the contribution of subjective probability to expected utility in games of partial information. Subjective probability corresponds to personal beliefs about the world that are difficult to quantify. A good example is the belief that members of the MLA are more likely to use *pen* to refer to a writing instrument than an enclosure for animals. I have no objective reason for this belief. In fact, I could well be wrong; perhaps some members of the MLA are deeply interested in animal husbandry, and thus the odds of their talking about animal enclosures are as high or higher than the odds of their discussing writing instruments.

My belief is based generally on what I believe about the MLA and the kind of people who are likely to attend its meetings. If you ask me to judge which seems more plausible, (17a) or (17b),

(17) a. A member of the MLA asks another member for something to write with by uttering, "Do you have a pen?"
b. A member of the MLA asks another member if she has an animal enclosure by uttering, "Do you have a pen?"

I have to say that I find (17a) more plausible than (17b). I've been a faculty member in a humanities department for some years now, and I can safely say that I can remember asking someone for something to write with, but I don't think I've ever discussed animal enclosures with any of my colleagues. Thus, when I'm asked to assess the probabilities of chance's move in figure 7.8, I assign far more value to p, the probability that the speaker intends a writing instrument, than to p', the probability that she intends an animal enclosure.

Of course, there's another way to think about the probabilities. If I could actually track the number of times I had heard *pen* with the interpretation of writing instrument and the number of times I had heard *pen* with the interpretation of animal enclosure, then I could work out an *objective measure* of the probability that *pen* is used in the sense of writing instrument. Let n be the number of times that I've heard *pen* in the sense of writing instrument and m be the number of times that I've heard it in the sense of animal enclosure. Then the probability that *pen* is being used in the sense of writing instrument is

$$\frac{n}{n+m}.$$

Notice that this probability would be based on experience of an individual. But, as the individual has more and more experience with the language, the objective probabilities that he assigns to word senses would tend to approximate the real probabilities of the population as a whole, although there would always be divergences between the probabilities assigned by an individual and the probabilities found in the entire population.

There is good evidence that people really do track the objective probabilities of word senses. For example, people are much quicker to recognize high-frequency words than low-frequency words, and they distinguish word senses based on frequencies. Since I work in a humanities department at a university, the objective probability I assign to *pen* as writing instrument is certainly higher than the probability I assign to the animal enclosure sense. It might take a cue from the environment—standing in the middle of a pig farm or perusing the latest issue of *Annals of the Congress of Pig Farmers*—to facilitate the animal enclosure reading for me.

This discussion is a bit abstract, so let's consider a concrete example. The sentence in (18) is a well-known example of what is called a garden path sentence; cues in the sentence lead readers down the garden path, as it were, to an incorrect analysis, which is then revealed by further information:

(18) The horse raced past the barn fell down.

I first encountered the sentence in (18) in a textbook when I was an undergraduate, and I couldn't make heads or tails of it. I concluded that it was utter nonsense; it took me several years before I figured out what the sentence was supposed to mean.

Compare the sentence in (18) with (19):

(19) The manuscript found in the trunk fell apart.

I assume that you found the sentence in (19) relatively easy to understand, compared to the one in (18). Nevertheless, (18) and (19) have nearly identical syntactic analyses. What makes (18) hard and (19) easy?

Consider paraphrases of (18) and (19) that are relatively easy to understand:

(20) a. The horse that was raced past the barn fell down.
b. The manuscript that was found in the trunk fell apart.

The sentences in (20) elucidate the intended meanings of (18) and (19). In fact, if we were constructing a game, we might contrast a choice between (19) and (20b).

Consider what goes wrong with (18) that makes it so hard to process. I have the strong intuition that the problem lies mainly (though not exclusively) with *race*. The problem is that I take *race* to be an intransitive verb—a verb that doesn't apply to a direct object—as in,

(21) The horse raced down the street.

This use of *race* seems more frequent than the transitive use of *race*:

(22) The jockey raced the horse down the street.

Furthermore, *the horse* in (18) makes for an excellent subject of *race* in the intransitive use. Notice that this preference is a bit weak; there is a transitive sense of *race* where *the horse* makes an excellent subject:

(23) The horse raced the tiger down the street.

Nevertheless, the frequency of intransitive *race* along with the goodness of fit of *the horse* as its subject reinforce each other. Thus, the intransitive sense of *race* is selected as having the highest utility, given the objective probabilities from word frequency integrated with preferences from the linguistic context. *The horse* can then combine as subject with the verb phrase *raced* to give a sentence:

(24) The horse raced.

The modifying phrase *past the barn* can easily be attached to this clause to give

(25) The horse raced past the barn.

But then the predicate *fell down* is encountered. This predicate wants a noun phrase subject like *the horse* (as in *the horse fell down*) but all we have to work with is the sentence in (25), which can't be integrated with *fell down* to make anything, at which point our sentence-processing equipment crashes and burns.

Compare this case with the more easily processed sentence in (19). Focus on the verb *found*, which is in a position analogous to *raced* in (18). In the case of *found* there is no more frequent intransitive sense. Instead, there must be a decision between whether *found* is simply the past tense form of *find*, as in

(26) The boy found a wallet in the street.

or whether *found* is a passive participle, as in

(27) The wallet was found in the street.

where the logical object of the verb occurs as its subject.

In the case at hand, notice that *the manuscript* makes for a very bad subject for *found*. Thus, the probability that *found* is a passive participle is increased. This means that when we combine *the manuscript* with *found*, the likeliest sense of the latter is as a modifier on *the manuscript*; in other words, we easily analyze the sequence

(28) the manuscript found in the trunk

as a noun phrase. When we encounter the predicate *fell apart*, we have no trouble integrating the two pieces.

Notice a sequence of increasing difficulty in the following examples because of the way in which the cues conflict:

(29) a. The bottle found in the room was an antique.
 b. The parasite found in the swamp was a new species.
 c. The boy found in the street was an orphan.
 d. The boy found in the street an object of inconceivable value.

As we move through the sequence, the initial noun phrase—*the bottle, the parasite, the boy*—becomes an increasingly plausible subject for *found*; in the last two cases, the absence of an object noun phrase immediately to the right of *found* increases the likelihood that the verb is part of a modifier on the noun phrase.

A simple example of a near garden path sentence illustrates the same point about the influence of context on the interpretation of words:

(30) The astronomer married the star.

A mild garden path occurs when the word *star* is encountered: *astronomer* has increased the subjective probability of the astronomical interpretation of *star*, raising its expected utility and, in the worst case, making this interpretation the Pareto-dominant equilibrium. The astronomical interpretation, however, is incompatible with the verb *married*, resulting in a feeling of semantic anomaly until the correct sense of *star* is retrieved.

The following is a simple economic model of lexical access.
If an ambiguous form is encountered,

1. all the different readings of the form are activated as choices;
2. each form is associated with its expected utility, where
a. the utility of the form is calculated along Gricean lines as discussed, and
b. the probability of the form is a function of its objective probability (that is, its lexical frequency) and cues from the linguistic environment;
3. the element(s) with the highest expected utility win and suppress items with lower expected utility;
4. if several elements are tied for first place, then one is chosen with probability $\frac{1}{n}$, where n is the number of forms in the tie.

This model is a game in the classical game-theoretic sense. Item (4) defines a mixed strategy, what to do when it isn't clear which form is correct. This is a game-theoretic interpretation of a fairly standard interactionist account of lexical access in psycholinguistics, so I can't claim any great insight on my part. The mathematics of game theory provide an elegant way of formulating this theory.

The theory of lexical access can be applied to the following case:

(31) The rich man the boats.

I've presented the example in (31) in classes where the students have uniformly rejected it as completely ungrammatical, but it is, in fact, a grammatical sentence of English.

The trick to working out (31) is to realize that *man* is used here as a verb meaning 'to provide the personnel to run something' (namely, the boats). The objective probability that *man* is a noun greatly exceeds that of its verb meaning, so the expected utility of *man* as a noun is far greater than the expected utility of *man* as a verb.

This expectation is reinforced by the presence of *rich* immediately to the left of *man*. Of course, *rich* could be taken as either an adjective, as in

(32) A rich life is not easy to attain.

or a noun, as in

(33) The rich have it easy.

with the objective probabilities favoring the adjective interpretation. Notice that the adjective interpretation of *rich* and the noun interpretation of *man* reinforce each other. The interpretation with the highest utility would then be a noun phrase, *the rich man*. Unfortunately, this cannot combine with the noun phrase *the boats*. By the way, the interpretation of *the rich man the boats* as a pair of noun phrases is sometimes correct, as in

(34) The rich man, the boats, and our treasure of gold doubloons were all lost in the storm.

Example (31) can be made into another example of perfect ambiguity with a small change:

(35) The rich man boats.

We would expect, given the word sense frequencies of *rich* and *man*, that the most accessible interpretation of (35) is the one where a particular rich man locomotes in a boat.

My wife, Jennifer, adores garden path sentences. When I told her about them, she began compulsively trying to make up garden paths. One night, around 3 a.m., she sat bolt upright in bed, looked at me and said,

(36) The moose head north.

Whereupon she fell back in bed and went to sleep.

Further Reading

Games of partial information were developed by Prashant Parikh in his dissertation in the late 1980s. His book *The Use of Language* (2001) lays out the formalism and discusses a number of applications, including implicature, illocutionary force, miscommunication, and jokes, as well as applications to esthetics, particularly visual representation. Parikh has extended games of partial information in a number of directions. See Parikh (2010) for an extension into truth-conditional semantics and Clark and Parikh (2007) for discourse anaphora, as well as chapter 8 of the present book.

As I noted, this chapter develops a Gricean account of lexical access. See Grice (1957; 1975), both of which are collected in Grice (1989). Grice's maxims have been the subject of an enormous amount of investigation since he proposed them. I have retained his original formulation, although the maxims can be reduced. See Horn (2001) and Levinson (2000) for a discussion.

My discussion of irony owes something to the work of Sally (2002; 2003), although I have not been able to include much discussion of his use of sympathy in the calculation of utility.

Many of the issues I touch on here were the subject of a vast commotion back in the 1980s involving modularity versus interactionism. Happily, we needn't worry too much about that debate; most of the issues are of no particular significance here. Two grand results are unassailable, I think. First, there is a huge effect for word frequency; there is a large literature on frequency in psycholinguistics. See, for example, Simpson (1994) or Norris (2006; 2009) and the references cited in those works. The other result involves semantic priming; again, the literature is vast, but the work of Ferrand and New (2003) is representative and accessible.

The garden path sentence *The horse raced past the barn fell down* is from Bever (1970), who may have coined the term *garden path sentence.* These sentences were central to the debate about modularity versus interactionism; I think Crain and Steedman (1985) were largely on the right path (heh). The account given here in terms of games of partial information is developed from Clark, Parikh, and Ryant (2007).

8 Two Examples: Pronouns and Politeness

In the past few chapters I've laid out the essentials of game-theoretic analyses of communication, including games of partial information. Of course, actually developing analyses is a different question. It's time to step through some problems and show how they might be analyzed using the techniques we've discussed.

Two examples are considered here. One is taken from Clark and Parikh (2007) on the interpretations of discourse pronouns: given a number of possible antecedents, when should a speaker use a pronoun with a reasonable expectation that the hearer will understand the pronoun as intended? This problem is clearly an example of a coordination game, one that is played frequently every day, by and large with great success.

The other example is suggested by the brief account of Gricean lexical access given in chapter 7. The present chapter investigates some aspects of the relation between politeness and conversational implicature, information that is suggested but not entailed by an utterance in a specific context. The topic of the relationship between politeness and Grice's conversational maxims is vast, of course; but we can map out some of the territory here.

The best way to understand game analyses is to do them. The two examples worked out here just touch on a vast area of study. I invite readers to work out examples on their own. There is a pleasure in doing so that can't be captured in reading about analyses, no matter how systematically examples are worked out.

Discourse Pronouns

Chapter 5 looked at quantifiers like *at least five tigers* from the point of view of their truth conditions and their effect on the discourse. The data were texts like the following:

(1) At least four monkeys teased a tiger. They were asking for trouble.

The pronoun *they* in the second sentence of (1) refers back to the monkeys that teased the tiger. The following rule was provisionally established:

(2) **(R.she)** (**Discourse Pronoun**)
Suppose the game is $G(S; M)$, and S contains the pronoun *she*. Then the current verifier may choose an entity, X, from the discourse model M_D and replace *she* by the name of that entity. The current falsifier may also choose an entity from M_D, Y. The game then continues as

$G(S'; M)$,

where S' is

$S[X/\text{she}]$ and X is female and if Y is female then $X = Y$.

The idea behind the rule (**R.she**) is that the verifier can pick someone (or something) as a witness to a sentence containing a pronoun as long as the witness exists in the discourse model, M_D; in other words, that entity must have been talked about before. Further, that entity must be unique in M_D; that is why the falsifier also gets to pick an entity. If that entity is also female, then the verifier will lose.

I noted in chapter 5 that the rule in (2) wouldn't really work because of the uniqueness requirement that the rule imposes. If there is more than one female entity in the discourse model, then the falsifier wins. The problem is that we use pronouns even when they do not have a unique referent in the discourse model:

(3) Mary slapped Susan. She$_1$ thought that she$_2$ had insulted her$_3$.

The fragment in (3) is fine as part of a discourse. Notice that for each of the pronouns in the second sentence, there are two possible candidates for their targets—Mary and Susan. My intuition is that Mary is the best candidate for *she*$_1$, Susan is the best candidate for *she*$_2$ and Mary is the best candidate for *her*$_3$.

Pronouns are an example of how language uses our ability to coordinate with each other to achieve communicative economy. A pronoun like *she* has almost no conventional content; its reference is determined by the context in which it occurs. Because we are able to coordinate our behavior, we can (usually) jointly fix the reference pronouns. By this I mean that the speaker uses a pronoun with the reasonable expectation that the hearer will be able to determine the intended reference of the pronoun, and the hearer is able to determine the reference of a pronoun because she can make assumptions about what the speaker intended the pronoun to refer to. In other words, the use of pronouns in discourse is a game; it involves strategic decision making on the part of both the speaker and the hearer.

In considering the use of pronouns in discourse, I put aside examples of coreference within a sentence, as in

(4) John told Bill that his pants were on fire.

In the example in (4), it seems that either *John* or *Bill* can be the antecedent—the intended referent—of the pronoun *his*. I've tested this by showing a small sample of speakers the written sentence (so they get no clues from intonation), and about half prefer *John* as the antecedent while the other half prefer *Bill*. In this case, with minimal information from context, the population appears to adopt a mixed strategy in resolving the reference of the pronoun *his*.

Things are quite different when coreference relations are compared across sentences. Consider the following text:

(5) An undercover cop was observing a suspect. He stayed in the shadows.

The first sentence in (5) sets up two discourse entities, the undercover cop and the suspect, while the second sentence uses a pronoun to pick one of them out. There is a distinct preference to take *undercover cop* as the referent for *he*—subjects have more salience than objects—so that it is the undercover cop, rather than the suspect, who stayed in the shadows.

This section, following Clark and Parikh (2007), develops some games that model the strategic decision making involved in resolving the reference of discourse pronouns. The idea is to model the use of discourse pronouns as following from the theory of rational behavior. The grammar does not say anything about the reference of discourse pronouns; in particular, there are no rules in the grammar that tell speakers and hearers how to interpret pronouns in a discourse. Instead, their reference follows from properties of the discourse context and the rational use of pronouns as a signaling device.

Both the speaker and the hearer have choices to make. As noted, the choice for the hearer is among the objects available in the discourse model. If two objects have been established in the discourse, the hearer needs to decide which of the objects is the target of the pronoun. Sometimes the hearer might be able to use gender information to make the decision. A hearer familiar with English names would know that *John* indicates a male, and *Mary* indicates a female:

(6) a. John was lecturing Mary. He thought she needed educating.
   b. John was lecturing Mary. She thought he was a blowhard.

So in the sentences in (6), the targets of the pronouns would be clear to that hearer because of gender. If, however, there were two discourse entities with the same gender, as in (3), then the hearer has a real choice to make.

Equally, the speaker must make the decision whether to use a pronoun or some other expression; the speaker might use a name or a definite description, for example. Compare the following texts with those in (5) and (6):

(7) a. An undercover cop was observing a suspect. The cop stayed in the shadows.
   b. An undercover cop was observing a suspect. The suspect stayed in the shadows.
   c. John was lecturing Mary. John thought Mary needed educating.
   d. John was lecturing Mary. Mary thought John was a blowhard.

So the speaker must decide whether to use a pronoun or some other expression.

The first step in an analysis of the problem is to represent the choices of the speaker and the hearer. To do this, let's take a concrete example, say example (5), where there is some ambiguity about which discourse entity the pronoun might refer to. The first sentence establishes two entities in the discourse model, the undercover cop and the suspect. Suppose the speaker intends to target either the undercover cop, e_1 in the discourse model, or the suspect, e_2. This can be modeled by letting nature make the first move, placing the speaker into s_1, the information state where he intends to refer to e_1, or into s_2, the state where he intends to refer to e_2. Of course, nature would move with some probability (figure 8.1). Assume that the probabilities, p and p', are disjoint and exhaustive, so that $p + p' = 1$.

Suppose the speaker is in state s_1, where he intends to refer to the undercover cop, discourse entity e_1. The speaker must decide whether to use a pronoun to refer to e_1 or devise some definite description:

(8) a. the undercover cop
   b. the detective
   c. the long arm of the law

Pronouns and Politeness

○ = nature
□ = speaker

Figure 8.1
Nature's Move in a Pronoun Game

Analogous reasoning applies if the speaker is in state s_2, where he wants to refer to the suspect; he can either use a pronoun or devise a description:

(9) a. the suspect
b. the shady character
c. the putative perpetrator
d. the alleged malefactor

Two things follow immediately. First, if the speaker uses a pronoun, the reference of the pronoun will be underdetermined for the hearer, who will have to decide whether the pronoun refers to e_1 or e_2. The players are therefore playing a game of partial information.

Second, if the speaker decides that a pronoun won't do, he will have to make some decisions about what description he should use to get the hearer to select the intended discourse entity. Let's denote these games by

(10) a. $G(s, h, e_1, c)$
b. $G(s, h, e_2, c)$

where (10a) is the game where the speaker, s, selects a description to signal discourse entity e_1 to the hearer h given a context c, and (10b) is the analogous game for signaling discourse entity e_2. There is a symbol for context, which could include information about common knowledge of the discourse, including the grammatical functions played by various phrases in earlier utterances.

Figure 8.2 shows the speaker's choices as well as the move by nature. Recall that information state s_1 embodies the speaker's intention to refer to e_1, the undercover cop, while state s_2 embodies the speaker's intention to pick out e_2, the suspect. The speaker can either use a pronoun to pick out the intended discourse entity or generate an alternative expression that will, in the context, uniquely pick out the desired entity. If the

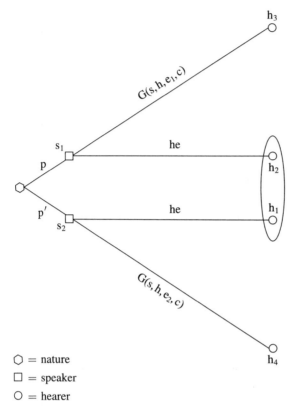

Figure 8.2
Speaker's Choices in a Discourse Pronoun Game

speaker uses a pronoun, then the hearer cannot be certain whether the speaker intended e_1 or e_2. This is represented by showing the hearer in the information set $\{h_1, h_2\}$.

Finally, the speaker could choose to generate a description that will clarify his intention; the branches for this choice are labeled with the corresponding games, but we assume that at the end of the game the speaker announces his choice and that this announcement will unambiguously denote the intended discourse entity. But recall from chapter 6 that something could go wrong because of bounds on the hearer's knowledge.

The hearer's choices (figure 8.3) take place after the speaker's action. The hearer can always choose between e_1 and e_2. In some cases, as when the hearer is in state h_3—the speaker has unambiguously announced his choice after playing game $G(s, h, e_1, c)$—the choice is obviously e_1, and

Pronouns and Politeness

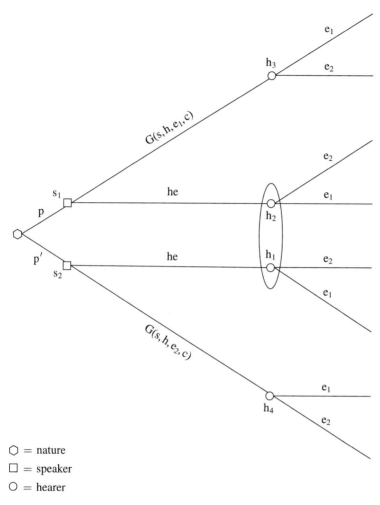

○ = nature
□ = speaker
○ = hearer

Figure 8.3
Hearer's Choices in a Discourse Pronoun Game

the hearer would only choose e_2 by accident. But in the information set $\{h_1, h_2\}$, the hearer is genuinely uncertain about which entity the speaker intends; she must use some strategic thinking to resolve the problem.

In order to solve the game, we must assign utilities to the outcomes of the players' actions. We need to work out the preferences of the players and then associate utilities in such a way as to preserve the ordering of the preferences. We could assign utilities in any way we like, which would rob the approach of interest; instead, we need to justify the way we assign utilities to the outcomes. This is where much of the theoretical work is done.

Let's take an explicitly economic approach and consider the problem in terms of production costs, risks, and the ultimate communication goals of the speaker and the hearer (Mayol and Clark 2010):

• Encoding costs. It is generally more costly to use longer expressions. This costs the speaker in terms of production effort and the hearer in terms of processing effort.
• Choice costs. It is generally more costly to use expressions with high conventional content; names and descriptions are costlier than pronouns. This cost involves the effort required to select among a host of possible names and descriptions that could be applied to an object. Here the speaker and hearer engage in a strategic choice that involves not only some direct costs but also potential risk if the selected expression is misunderstood.
• Context factors. It is cheaper to refer to a more prominent element with a pronoun. It is correspondingly more marked (hence, costlier) to refer to a more prominent element with a description or name when a pronoun could be used. Here the speaker and hearer are presumably using some notion of salience to guide their choice.
• Communication factors. Successful communication, where the speaker transmits the intended message successfully, is highly preferred. Miscommunication, where the hearer misinterprets the speaker's intended message, is strongly penalized.

Let's consider the impact of each factor on ranking the outcomes.

Encoding costs represent the idea that the longer an expression is, the more effort is required to produce and process it. This increase in effort is a cost that is deducted from the overall utility of an expression. In general, (short) pronouns are preferred over longer expressions like descriptions:

Pronoun > Description.

The idea that there is pressure to select shorter expressions is not new; one finds it in many places. For instance, Grice's conversational maxim "Be brief (avoid unnecessary prolixity)" is one manifestation of the preference for shorter expressions. Other forces may be at work, though, which might lead to a preference for a longer expression in some cases.

Choice costs express the idea that generating a novel description or using a name entails some extra costs. To generate a novel description, the speaker will have to try to select a description that he knows (or, at least, has reason to believe) will signal the intended referent to the hearer. This is not a trivial matter, since failure to select a workable description will result in miscommunication. The speaker will have to balance the benefit of successful communication against the cost of adding information via the description. For example,

(11) John broke up with Mary. The idiot doesn't know what's good for him.

involves using *the idiot* to refer to John and thus provides the hearer with information about the speaker's attitude toward John. Notice, though, that *the idiot* could also refer to Mary:

(12) John left Mary. The idiot kept cheating on him.

So there is some risk involved in using a novel description.

Equally, using a name also risks miscommunication if, for example, the hearer does not know the name of the discourse entity. Thus, the speaker must solve a strategic decision problem even in using a proper name, and this entails some costs. Pronouns, on the other hand, require only gender, number, and information available from the context.

Context factors emphasize the importance of salience in determining the reference of pronouns. I've mentioned the importance of focal points in natural language (see chapter 9 for details). Suppose the subject of the sentence is a focal point, all else being equal. This means that there should be a preference for the subject over other grammatical functions:

Subject > Other.

In fact, *centering theory* holds that there is a general prominence hierarchy of grammatical functions:

Subject > Indirect object > Direct object > Other.

Prominence closely tracks the distribution of animate noun phrases, so there may be a complex interaction between semantic properties and grammatical function in working out focal points.

The focal point is not always determined by grammatical function. If an individual is sufficiently salient in the local environment, she can become a focal point and is a good target for a pronoun. For example, suppose a woman collapses in the street; a bystander might felicitously say,

(13) She needs a doctor.

The pronoun clearly denotes the woman who collapsed.

The focal point might shift depending on lexical properties of the verb or discourse connectives.

Communication factors are an important part of the economics of speech. If the speaker fails to signal the intended meaning to the hearer, there is a potential for misunderstanding and a delay while the participants go back to repair the problem. The participants in the conversation might not notice the miscommunication; in fact, unremarked miscommunications may be quite frequent (see chapter 6). Nevertheless, the threat of a derailment due to miscommunication is sufficiently dire to warrant a penalty, and both the speaker and the hearer would be well-advised to keep the problem in mind.

Figure 8.4 shows a pronoun game for example (5), which is repeated here:

(14) An undercover cop was observing a suspect. He stayed in the shadows.

Recall that in state s_1 the speaker intends to pick out the undercover cop. The pronoun option receives the highest utility here; since *undercover cop* is a subject, the utility is (3, 3) as opposed to (2, 2) for *suspect* (the choice dominated by state s_2), which is an object. This assignment is in accord with the context factors; all else being equal, the subject is the focal point.

The pronoun option, when successful, has higher utility than the description or name option, as is consistent with both encoding costs (pronouns are shorter than descriptions) and choice costs (pronouns presuppose less than descriptions or names). Finally, there is an asymmetry between choosing a description for something that is prominent—the subject of the preceding clause—versus a description for a nonsubject; there is a slight penalty for using a description for the prominent element. Thus, the game in figure 8.4 accurately reflects the preferences that have been outlined.

I've included, from states h_3 and h_4 in the figure, the possibility that the hearer selects the incorrect discourse entity despite the (presumably) unambiguous nature of the description or name that the speaker has chosen. A rational agent would never make this choice, since a utility of 1 is always preferable to a penalty of -2. These branches can be ignored if the reader finds them distracting.

Let's work out a strategy profile for the players. Assume that nature is as likely to put the speaker into state s_1 as into state s_2, so $p = p' = 0.5$. In order to calculate expected utility, the utilities dominated by s_1 must

Pronouns and Politeness

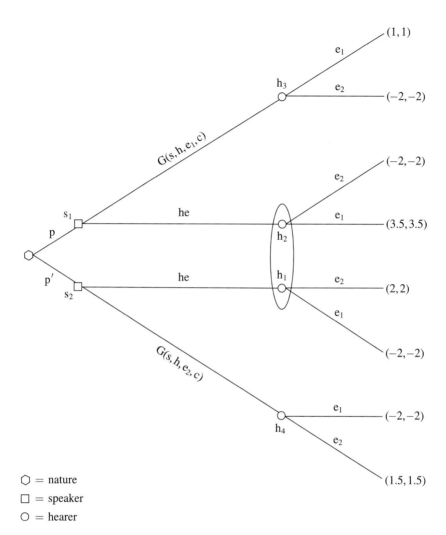

○ = nature
□ = speaker
○ = hearer

Figure 8.4
Simple Pronoun Game

be multiplied by p, and the utilities dominated by s_2 must be multiplied by p'.

We can use backward induction. Let's start from state h_3. The hearer would be irrational to choose e_2 and suffer a penalty; clearly she will prefer to choose e_1 for a net gain. Thus,

(h_3, e_1).

Analogous reasoning from state h_4 gives

(h_4, e_2).

Now consider what happens in the information set $\{h_1, h_2\}$. By hypothesis, half the time the speaker intends e_1, and half the time the speaker intends e_2. From the expect utilities, there is a slight preference for selecting e_1; we would on average expect to get 0.75 for that choice as opposed to an average of 0 for choosing e_2. Therefore,

$(\{h_1, h_2\}, e_1)$.

The prominence of the subject gives it an edge as the target of a pronoun in the next sentence.

Now consider the speaker. Suppose nature has placed him into state s_1; he can either generate a description or use a pronoun. He knows that if he uses a pronoun, the hearer will select e_1. Since the speaker's utility for the pronoun option is greater than his utility for the description, a rational player will pick the pronoun:

(s_1, he).

Suppose, though, that nature has placed the speaker into state s_2, where he wants to denote the suspect, which was the object of the preceding clause. If the speaker uses the pronoun, he has every reason to believe that the hearer will select e_1. Miscommunication! Since the speaker prefers a positive utility, he has every reason to choose to generate a new expression rather than use a pronoun, so

$(s_2, G(s, h, e_2, c))$.

We can now assemble the following strategy profile for a one-pronoun game:

$\{(s_1, he), (s_2, G(s, h, e_2, c)), (\{h_1, h_2\}, e_1), (h_3, e_1), (h_4, e_2)\}$.

This strategy profile suggests that, all else being equal, the speaker will tend to use a pronoun to pick out the subject of the preceding sentence and generate a description to pick out any other element in the discourse model. The sentence in (14) should be rated as natural, with *he* denoting the undercover cop. The interpretation of the pronoun as the suspect should be considered unnatural.

If the speaker wishes to pick out the suspect, then he will generate a description; the most accessible description would be to repeat the predicate used to introduce the entity in the previous sentence:

(15) An undercover cop was observing a suspect. The suspect was behaving in an odd manner.

Notice that using a description to pick out the undercover cop should be considered relatively unnatural:

(16) An undercover cop was observing a suspect. The (undercover) cop stayed in the shadows.

Indeed, the sentence in (16) seems overly stiff and didactic, verging on the impolite in its pedantry.

We've seen that the game approach works for a simple case of one pronoun. The utilities associated with the outcomes reflect the preferences of the players; any assignment of numbers that reflects these preferences should preserve the result. Notice that the real work lay in developing the theory of preferences given by the production costs.

This game approach can be scaled up to a more complex case to account for the flexibility seen in actual texts and to demonstrate how different factors in the context can apparently change intuition about the interpretation of discourse pronouns.

Consider a case with two discourse entities that could serve as potential targets for a pronoun, and a sentence containing two pronouns. The question is whether we can account for preferences in the relation between pronouns and discourse entities.

(17) Mary tripped Susan. She hadn't noticed her.

In principle, there are two interpretations available for the second sentence in (17). Mary might not have noticed Susan, causing her to trip, or Susan might not have noticed Mary, resulting in the trip. Either interpretation is possible in the absence of other information. There is a preference for the first interpretation, however: *she* denotes Mary (the subject of the first sentence), and *her* denotes Susan.

Nature can place the speaker into state s_1, where he wants to continue talking about Mary, so an element denoting Susan would be in the object position; or nature can place him into state s_2, where he wants to talk about Susan, in which case an element denoting Mary would be in the

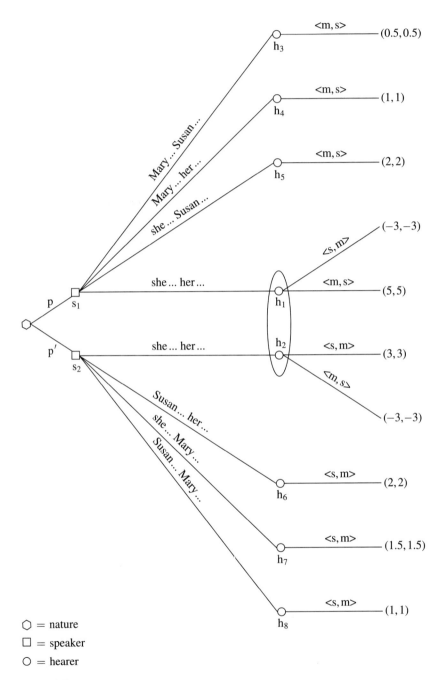

Figure 8.5
Pronoun Game with Two Pronouns

object position. Suppose that the only question for the speaker is whether to use a pronoun or a proper noun to denote Mary and Susan.

Figure 8.5 shows the game tree with the obviously bad choices suppressed. At each state s_1 or s_2, the speaker has four options to choose from, depending on whether he uses a name or a pronoun for the two discourse elements. Consider first the ranking of his choices from state s_1. If he could, he would most like to refer to both discourse entities with pronouns:

(18) She hadn't noticed her.

This option would give him the greatest utility; of course, it also carries the most risk, since the hearer might make the wrong choice. His next preference would be to refer to Mary, the subject of the preceding sentence, with a pronoun and Susan with the name *Susan*:

(19) She hadn't noticed Susan.

This uses a pronoun for a focal element but a full name for another element, so its utility is not as high as the two-pronoun case, but it also carries less risk.

The case where the speaker uses the name *Mary* and a pronoun for Susan,

(20) Mary hadn't noticed her.

is relatively marked; it uses a pronoun for a nonfocal element, the object of the preceding sentence, and a full name for a focal element. The utility is therefore relatively low.

The lowest utility is associated with the use of two names:

(21) Mary hadn't noticed Susan.

Contentful forms are used everywhere, including for a focal point.

Now consider the options under the other state, s_2, where the intended subject is Susan. The highest utility is associated with the two-pronoun form; the utility is not as high as the case where Mary is the subject, since there is a preference to keep the focal point in subject position.

The next option is to use the name for Susan and a pronoun for Mary. This has the advantage that it uses a pronoun for the focal element, which is now in object position:

(22) Susan hadn't noticed her.

Somewhat less good is the case where a pronoun is used to pick out the nonfocal element, Susan, and a name is used for the focal element:

(23) She hadn't noticed Mary.

Finally, the case where both discourse entities are denoted by names is least preferred. Notice that the utilities match the preferences according to the production costs.

Let's construct a strategy profile for the players. As before, assume that nature is equally likely to place the speaker into state s_1, where he intends to keep Mary as the subject of the next utterance and Susan as the object, or into state s_2, with the opposite assignment of grammatical functions.

From the hearer's point of view, most of the information states have an obvious solution:

$\{(h_3, \langle m, s \rangle), (h_4, \langle m, s \rangle), (h_5, \langle m, s \rangle), (h_6, \langle s, m \rangle), (h_7, \langle s, m \rangle), (h_8, \langle s, m \rangle)\}.$

These are just the choices with the highest utility. The interesting question is what to do at the information set $\{h_1, h_2\}$. Although the hearer has equal chances of being in h_1 or h_2, according to the expected utilities for the information state the hearer should have a preference for selecting the pair $\langle m, s \rangle$—Mary is the target of the subject pronoun and Susan is the target of the object pronoun—since that option yields a positive expected utility:

$(\{h_1, h_2\}, \langle m, s \rangle).$

Now let's consider the speaker's choices. Suppose nature places him into state s_1. His best choice is to use the pronouns,

(24) She hadn't noticed her.

since that option nets him the highest utility.

What if nature places him into state s_2? In this case, his best option is to use the name to denote Susan and the pronoun to denote Mary:

(25) Susan hadn't noticed her.

So we can complete the strategy profile:

$\{(s_1, she \ldots her \ldots), (s_2, Susan \ldots her \ldots), (\{h_1, h_2\}, \langle m, s \rangle), (h_3, \langle m, s \rangle), (h_4, \langle m, s \rangle), (h_5, \langle m, s \rangle), (h_6, \langle s, m \rangle), (h_7, \langle s, m \rangle), (h_8, \langle s, m \rangle)\}.$

That is, the following texts are considered natural:

(26) a. Mary tripped Susan. She hadn't noticed her.
    she = Mary; *her* = Susan.
   b. Mary tripped Susan. Susan hadn't noticed her.
    her = Mary.

The other combinations seem somewhat less natural. For example,

(27) Mary tripped Susan. She hadn't noticed Susan.

has a slightly stilted air because of the repetition of *Susan* in object position.

The analysis of production costs for pronouns and nonpronominal noun phrases sits outside the grammar. It is a theory of how rational agents will use the output of the grammar to signal their intended meaning. The theory provides a workable result simply by computing the Pareto-dominant equilibrium.

We need to test and extend the result. So far, it has been assumed that the speaker is as likely to continue talking about the subject as to change topic to the object. If the speaker is more likely to talk about the subject, nothing in the analysis changes. But what happens if the speaker is more likely to change topic and talk about the object? Compare the following two texts:

(28) a. John called Bill a republican. Then he insulted him.
 b. John called Bill a republican. Then *he* insulted *him*.

Read with a flat intonation, (28a) describes a sequence of events: first John called Bill a republican, and then John insulted Bill. This is exactly as expected, given the game in figure 8.5, which is isomorphic to (28a); the subject of the preceding sentence, the focal point, is the likeliest target for the pronoun *he*, leaving the object as the target for *him*.

The speaker can signal his intention to change topic by using contrastive stress; the stress in (28b) suggests that first John insulted Bill by calling him a republican, and then Bill insulted John back.

In terms of the game, contrastive stress indicates that p', the probability that the speaker intends to speak about the preceding object, is greater than p. The utilities in the game tree in figure 8.5 can be used to calculate the result. Since contrastive stress indicates that $p' > p$, let $p' = 0.8$ and $p = 0.2$. Now, when the hearer interprets *he insulted him*, the expected utility of choosing *Bill* for *he* and *John* for *him* dominates. With a signal that the speaker intends to change topic, there is a pop-out effect; the Pareto-dominant equilibrium changes.

This is an interesting result. It shows that we can keep the underlying choices and utilities constant and simply vary the probabilities that the speaker and hearer assign to the information states. There are a variety of ways that the participants' estimates of the probabilities can be manipulated. One way is by contrastive stress. Another way is to use discourse connectives. Compare the following two texts:

(29) a. Mary insulted Susan. Then she slapped her.
   b. Mary insulted Susan. So she slapped her.

In (29a) *then* signals that the speaker is describing a sequence of events; nothing makes the object a better focal point than the subject. In (29b) the lexical semantics of *so* indicates that the preceding clause is a justification for the action of the second sentence; Susan slaps Mary because Mary insulted her.

So the semantics of items can affect the assessment of probability about the speaker's intentions. Compare the following:

(30) a. John swindled Bill. He's a rather shady character.
   b. John swindled Bill. He should be more prudent.

The text in (30a) behaves as expected, given the basic game; *he* targets *John*, the subject of the preceding sentence. In (30b) the semantics of the predicate, *be more prudent*, biases the interpretation toward assigning *Bill* as the target of *he*; in terms of the game, it increases the subjective probability of s_2.

Let's consider the contribution of focal points a little more carefully. Consider the following text:

(31) John's pasta sauce got on Bill's jacket. He didn't notice.

As things stand, we make no predictions about the interpretation of the pronoun *he* in the second sentence of (31). Neither *John* nor *Bill* occupies a sufficiently prominent grammatical function to be a focal point. A small survey of speakers indicates that when forced to choose, some pick *John* as the target of the pronoun and some pick *Bill*. Thus, we seem to have a case of a mixed strategy Nash equilibrium.

Things change, however, if we can make *John* or *Bill* a focal point. Consider the following:

(32) Bill always has his head in the clouds. He doesn't see what's right in front of him. Last night, John's pasta sauce got on Bill's jacket. He didn't notice.

Here, *he* in the last sentence is clearly intended to pick out *Bill*. This is because the text is about Bill, so *Bill* is the focal point of the set {John, Bill}.

The interpretation changes if we make *John* the focal point:

(33) John is unbelievably careless. When he cooks his pasta sauce, it gets all over everything. Last night, John's pasta sauce got on Bill's jacket. He didn't notice.

In (33) the focal point of the set is *John*, not *Bill*. As expected, *John* is a better target for the pronoun in the final sentence than *Bill*. Being the focal element of the set of available entities in the discourse model should raise the subjective probability that that element is the target of a pronoun.

This approach to discourse pronouns can be extended by studying sequential games. The reference of a discourse pronoun is resolved by attending to grammatical focal points; the most prominent discourse entity is the likeliest target for a pronoun. This prominence is carried forward through a text, being either maintained or downgraded. That is, a speaker might choose to change the focal point while telling a story:

(34) Mary slapped Susan. She has a short temper and Susan was being annoying. She has always been socially inept.

The story in (34) involves a change in focal point midway through from *Mary* to *Susan*. The way that people manage the organization of narratives brings up a new area for game theory. The speaker might make higher-order decisions, like changing the focal point of a text, that might be locally costly in terms of expected utility but globally allow for a greater return of utility. Of course, exploring the strategic structure of narratives would venture far outside the scope of the present work.

The game-theoretic analysis of discourse pronouns strongly suggests that the use of pronouns in discourse is an example of rational behavior. Thus, we don't need to add a specific component to the grammar to handle the interpretation of discourse pronouns.

There is a great deal more to say about discourse anaphora. For example, there is a choice in many of these games between pronouns and definite descriptions. This suggests that there is a functional similarity between pronouns and definite descriptions. In general, it seems that a definite description is used to pick out an entity in the discourse model when that entity is not sufficiently salient to be the target of a pronoun. Exploring this analogy between definite descriptions and pronouns is best left for future work.

Politeness, Power, and Implicature

It is well known in my department that I'm very bad about schedules. I tend to forget what time it is and often don't pay attention to calendars. The administrative assistant, Amy, knows this particularly well. One day, when she knew that I had a grant meeting at 4 p.m., she stopped by my office and said,

(35) It's five minutes to four.

and then walked on. Of course, when I heard this, I got up immediately and went to the meeting, which was in a nearby building.

The example illustrates conversational implicature. Grice's maxims and implicature as well as a Gricean theory of lexical access were discussed in chapter 7. This section explores in more detail how to treat Grice's maxims game-theoretically, and it relates the conversational maxims to a theory of politeness.

What Amy literally said was merely a statement of the time of day, but she implicated that I should go to my meeting. Notice that although I might claim that she really wanted me to go to the meeting, she can always deny that she meant that. She might claim that she only intended to tell me the time. Conversational implicatures, unlike entailments, can almost always be canceled. Amy might, for example, have said,

(36) It's five minutes to four, but I don't mean to tell you to go to the meeting.

in which case she explicitly cancels the implicature.

I assume that my interlocutor, in this case, Amy, is a cooperative person. Being cooperative means that she tries as much as possible to obey Grice's conversational maxims (figure 8.6). In particular, she tries, in the main, to obey the maxim "Relation: Be relevant."

Seen from my perspective, Amy has violated the maxim of relevance; dropping by my office and announcing the time out of the blue is hardly relevant to anything. "But," I might say, "wait a minute. At 4 p.m. I have an important meeting. If I leave now, then I'll just make the meeting." So I can repair Amy's apparent violation of a conversational maxim by assuming that she's really telling me to go to the meeting.[1]

Implicatures often have the property that in order to account for the speaker's behavior, the hearer must recognize that the speaker's apparent violation of one or more of the maxims can be repaired if the speaker was trying to convey some implicated message. In particular, I am faced with the following decision problem:

(37) Should I leave for my meeting now?

Amy's utterance supplies me with a piece of information that helps me solve the decision problem in (37). So what she said was actually relevant.

Here's the puzzle, though: why didn't Amy just come out directly and say,

(38) Robin! Go to your meeting!

> Maxims supporting the cooperative principle (Grice 1975):
>
> **Quantity**
>
> 1. Make your contribution as informative as is required (for the current purposes of the exchange).
> 2. Do not make your contribution more informative than is required.
>
> **Quality**
>
> Try to make your contribution one that is true.
>
> 1. Do not say what you believe to be false.
> 2. Do not say that for which you lack adequate evidence.
>
> **Relation**
>
> Be relevant.
>
> **Manner**
>
> Be perspicuous.
>
> 1. Avoid obscurity of expression.
> 2. Avoid ambiguity.
> 3. Be brief (avoid unnecessary prolixity).
> 4. Be orderly.

Figure 8.6
Grice's Conversational Maxims

instead of being indirect and risking the possibility that I would fail to make the connection? Since I'm so notoriously absent-minded, shouldn't she just assume that I'll fail to draw the inference? This is the problem I want to address here. Why should speakers use indirection, in the form of conversational implicatures, to achieve their means when they could just as well be direct, get their message across, and avoid the risk of being misunderstood? Of course, the answer involves politeness. This section explores how politeness relates to strategic interaction and the conversational maxims. The example at hand can be used to lay out some of the principles, although a full treatment would go well beyond the scope of this book.

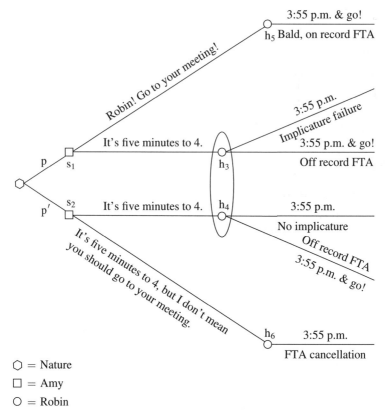

Figure 8.7
Implicature and Strategic Choice

Consider the question of common knowledge. Amy and I both know that I have to solve the decision problem in (37). Amy thinks that if she supplies me with the relevant piece of information, then I will solve the problem and go to my meeting. Amy and I have the choices shown in figure 8.7. This is the game tree minus the utilities, which are worked out later.

Assume that nature determines whether Amy wants to tell me the time in order to get me to go to the meeting, represented as state s_1; or whether she simply wants to let me know the time and doesn't care whether I go to my meeting, represented as state s_2. Nature decides which state Amy is in according to some probability mass function P, where $P\{s_1\}$ is p, and $P\{s_2\}$ is p'. For the moment, assume that $p + p' = 1$.

Assuming that Amy is in s_1, she intends to signal not just the time but the desirability of my leaving for the meeting. She therefore must decide whether to use a direct command, "Robin! Go to your meeting!" or to use indirection, "It's five minutes to four." In the latter case, she hopes that I will get both the literal meaning of the utterance and the implicature that I should get going. Equally, if Amy is in s_2, then she intends merely to tell me the time. In that case, if she tells me it's five minutes to four, I might not only get the literal content of her statement but also draw the conclusion that I should get going. This may not seem like much of a risk for Amy, but if she's worried I might draw that conclusion, she can always cancel the implicature explicitly. She might not want me to make the inference for fear that I might take offense; the implication that she is commanding me to go to a meeting might be taken as violating the power asymmetry in our relationship.

Now, if Amy tells me explicitly, "Go to your meeting!", then I have little doubt as to her communicative intent. If she uses the more indirect method of announcing the time, then she is relying on me to be cooperative and draw the correct inference. As Grice (1975) puts it, I must take the uptake. The strategic interaction here is reminiscent of a stag hunt game (see chapter 4). If I'm cooperative, then both Amy and I get some payoff; Amy accomplishes her goal of getting me to the meeting, and I accomplish my goal of attending the meeting. Further—and this is the point that needs to be explored—Amy will have succeeded in her task *politely*; she will not have impinged on my sense of social autonomy.

If I don't cooperate, then we fail to get the main prize, although Amy has succeeded in telling me the time, and I've at least understood what time it is. However, in order to accomplish her goal of getting me to the meeting on time, Amy might have to turn to more drastic means, which might involve an expenditure of social potential on both our parts.

These considerations bring us to the next step in the game analysis, the assignment of utilities to the outcomes of the choices. Here, we must directly address the puzzle of why Amy would use the risky strategy of indirectness—where miscommunication becomes a real possibility—instead of being more direct and thus guaranteeing successful communication.

Given that, on balance, a great deal of utility is associated with successful communication, I can only assume that some other factor outweighs the guarantee of success. Following interesting work by Goffman (1959) and Brown and Levinson (1987), I'd like to suggest that the crucial factor is *face*. Face is a fundamental force in the social world. Goffman defines face as the "positive social value a person carries." Face is an active construction, the image of the self that a person attempts to project to other members of society. With face, I can impact my social lot; without face, I am socially impotent, unable to direct or influence the course of social action around me.

Notice that face is a social construction. I am interested in presenting and maintaining a certain type of face, as a person of social worth with a certain amount of power and certain rights and privileges. When I lose face, I feel a sense of embarrassment, a loss of control over the situation I find myself in; I must bear up until I can recover some face and again am able to project a positive image of myself.

Under normal circumstances, I am also interested in maintaining the face of those around me. If I am upset by my own loss of face, I am equally disconcerted when someone around me loses face. Who has not felt a sense of embarrassment at someone else's embarrassment? This feeling, often taken to be simple empathy, is actually quite complex, since it engages discomfort over the temporary collapse of social order, a kind of local anomie. Face sits at the very heart of our cooperative behavior. Chapter 4 briefly took up the question of the origins and evolution of conventions and cooperation. I believe that an understanding of face is fundamental to an understanding of cooperation and the evolution of convention.

But more particularly, I would argue that we cannot really understand Grice's conversational maxims unless we understand the strategic role of face in our ordinary interactions. Seen in this light, understanding implicature becomes central to our broader understanding of cooperative behavior.

The idea that face is central to our understanding of implicature goes back at least to Brown and Levinson (1987); game theory can be used to augment their study. They argue that understanding politeness involves understanding rationality and face. Game theory gives us a theory of rational behavior within which we can embed a theory of face.

Face can be divided into two sorts:

• Negative face. This is freedom of action and freedom from imposition, the right to personal space. Every competent adult member of a society wants to be able to act without being impeded by others.
• Positive face. This is the self-image claimed by social agents. In particular, individuals want to be viewed positively and want their wants to be viewed as desirable.

Positive face, then, involves the image that an individual projects and its positive perception by others, while negative face involves the ability to act independently in the world.

Of course, things do not always go smoothly in the social world, so people are sometimes compelled to threaten another person's face by performing a *face-threatening act* (FTA). For example, suppose the dean of undergraduates at a fictional university has forgotten his wallet. He needs some cash but doesn't have his debit card, so he wanders around campus looking for a likely source of a loan. He first sees a professor emeritus doddering down the steps of the library. "I won't ask him," thinks the dean. "He has the money, but he won't take kindly to my asking for a loan. He'll probably tell the dean of arts and sciences I'm broke, and that would be embarrassing!"

Next, he sees the chair of the linguistics department. "Not a likely victim," thinks the dean. "He's too cheap, and anyway he never carries money with him." Finally, the dean sees an assistant professor from the business school. "Ah!" he thinks. "This is just right. He's from the business school, and they always have money. He's also an assistant professor; he'll be too nervous about tenure to turn me down."

The dean closes in. "Howya doin', pal?" he says to the assistant professor, who cowers and tries to look friendly. "Say, could you float me a loan of 20 bucks?" Here's the FTA. Notice, first, that the dean has carefully selected someone whose face he could threaten with little consequence.

Next, consider the request from the point of view of the victim. On the one hand, he likes to project the image of a generous and accommodating person who's always willing to help someone out. This is his positive face, and if he fails to lend the dean twenty dollars, he's afraid of projecting an image that is inconsistent with his positive face, particularly when the supplicant is higher on the ladder than he is. Thus, the dean has threatened the assistant professor's positive face by asking for a loan.

Equally, the assistant professor had hoped to use the twenty dollars to help pay for dinner at a local Ethiopian restaurant that evening. The sacrifice of the money to satisfy the dean's request limits his action; if he says no, he loses face, but if he agrees, then he's constrained to give the dean the money. The dean has therefore threatened the assistant professor's negative face as well.

Face-threatening acts come in two varieties; they can threaten either positive face or negative face. Correspondingly, there are acts that are intended to provide redress for FTAs: negative politeness, which is intended to ameliorate a threat to negative face, and positive politeness, which redresses the person's positive face.

Positive politeness is usually centered on the assurance that the threatener actually does want what the threatened person wants, thus sanctioning the desirability of their wants. Negative politeness is avoidance-based and often involves self-abasement on the part of the threatener. The dean of undergraduates, for example, could have softened his request by adding some negative politeness to his request,

(39) Look, I feel like a real jerk for asking, but I've stupidly got myself in a jam; could you lend me twenty dollars?

which would have lightened the sting a bit with a little kowtowing.

Brown and Levinson (1987) give the typology of FTAs (figure 8.8). The diagram resembles a decision tree, but it is really a hierarchical classification of face-threatening acts. Notice that the terminal nodes are numbered; the higher the number, the less the cost of performing the action. I could just choose not to do the FTA; then whatever task I might have wanted to accomplish by performing the FTA might not happen, but I will not incur the cost of performing an FTA.

Supposing that I decide to do the FTA. I can choose to go on record with the FTA, meaning that it will be common knowledge that I've performed the FTA; or I can go off record. For example, the dean of undergraduates could have said something like,

(40) (*Patting himself in search of his wallet*) Oh no! I left my wallet at home so I'm broke and I don't have my debit card.

hoping all the while that the assistant professor will charitably step in and offer him some money. We are here in the realm of implicature, as Brown and Levinson demonstrate; the dean hopes that the assistant professor will get the uptake and offer money. On the other hand, the dean risks not only that the assistant professor will not get the point, but also that even if he does offer money, he might not offer enough; the dean needs twenty dollars and the assistant professor might offer him a fiver.

The dean might have no choice but to go on record with his FTA. He must then decide whether to perform the action baldly (without any redressive action) or do something to save the assistant professor's face. The potentially most costly action would be to do the action baldly, as the low number associated with this option in figure 8.8 indicates. This would be a case where the dean simply demands twenty dollars in the manner of a schoolyard bully demanding someone's lunch money.

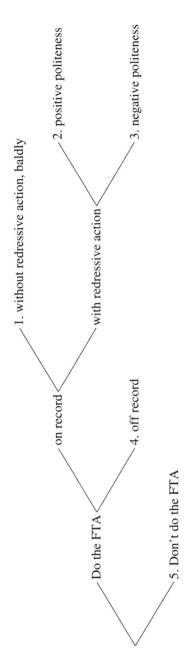

Figure 8.8
Typology of Face-Threatening Acts (From Brown and Levinson 1987)

In the polite society of the fictional university, such behavior is beyond the pale. The dean is likely to offer some redress in the form of politeness. The least costly would be some form of negative politeness. The transgressor might apologize for imposing on the victim, hedge the illocutionary force of the request (a command) by disguising it as a question, or engage in some nonverbal redress like assuming an apologetic posture. Finally, the dean could engage in some act of positive politeness by implying that the assistant professor's wants are indeed his (the dean's) wants as well; he might try to imply that the assistant is part of the dean's social group, that they're friends, and so on. Obviously, there is more than enough fuel here for lifetimes of research on strategic interaction.

The dean in this story would have to do a certain amount of work to decide whether or not to do the FTA; if so, whether to do it on record; if on record, whether to perform some redressive action; and so on. As noted, he chose not to ask an emeritus for a loan—also part of the decision work. Brown and Levinson give an equation that points to a number of factors:[2]

$$\text{Weight(act)} = \text{Distance(Speaker, Hearer)} + \text{Power(Hearer, Speaker)} + \text{Gravity(act)}.$$

That is, the weight associated with performing a face-threatening act is related to the social distance between the speaker and the hearer, the power relationship between the two, and the relative gravity that the culture associates with the act. Thus, it is usually costlier to perform a face-threatening act if there is a great deal of social distance between the participants; I'm less likely to ask a complete stranger to lend me twenty dollars than someone I know well. Equally, the power relationship between the participants is crucial in determining the weightiness of an FTA. Someone higher in the social ranking has an easier time imposing on someone lower, and someone lower in the social ranking would find it harder. Thus, the dean found it easier to ask an assistant professor for a loan than to ask an emeritus or a chair.

Finally, the last term is an estimate of the gravity of performing the act. How much of an imposition is the act, according to the culture? Asking someone for the time is not as much of an imposition as asking for a loan. It would be absolutely bizarre to hear the following:

(41) I'm really sorry and I feel like a jerk for imposing on you. But I left my watch at home and my cell phone battery died. Could you possibly find it in your heart to tell me the time?

In this case, the politeness formulas far outweigh the gravity of the act. The cost of producing so much blather is sufficiently high that the hearer could only conclude that the speaker was trying to signal something by saying so much. The obvious inference would be that the speaker was somehow mocking the hearer or disparaging the hearer's generosity. Now we have come back full circle to conversational implicature. Example (41) violates Grice's manner maxim, since the speaker is inappropriately prolix. The hearer, in order to account for such outlandish behavior, must form a hypothesis that explains why the speaker would behave so strangely. Fisticuffs would no doubt ensue.

This brief outline of a theory of politeness can now be folded into the game-theoretic account of implicature (and Amy's act of implicating that I should go to my meeting). The limited means available are the action choices of the participants, the probabilities associated with certain states, the outcomes of the actions, and the utilities associated with those outcomes. The only plausible method to account for face and implicature is to work out their impact on the preferences of the speaker and the hearer.

Consider the conversational maxims in figure 8.6. Grice intended them as a model of ideal conversation, that is, as long as things were flowing along nicely, speakers would obey all the maxims. This suggests that a small increase in utility should be associated with each maxim. All else being equal, an utterance that obeys all the maxims will have the highest utility.

Of course, an utterance that violates a maxim carries an implicature with it. Intuitively, it should be the case that the additional information carried by the implicature adds to the utility of the utterance, given the local model and choices available to the speaker and hearer. There may be a number of ways that an utterance might gain utility while violating a maxim, but here I focus on face and politeness.

Recall that I am in my office facing the decision problem of whether or not to get up and go to the meeting. Amy, the administrative assistant, helps me out by telling me that it's five minutes to four. She faces a choice between telling me the time or telling me outright to go to the meeting. Notice that both options solve the decision problem for me, either by giving me the information to solve the decision problem or by giving me the solution itself.

However, telling me to go to the meeting is a face-threatening act; since it places constraints on my freedom of action, it clearly threatens my negative face. Thus, although it obeys all the conversational maxims, we must deduct the cost of an on-record FTA without redress. The act is made somewhat worse by the fact that there is a power asymmetry in our relationship; as a faculty member, I have somewhat more power than an administrative assistant. The cost of a bald FTA in this power asymmetry, particularly one involving negative face, is sufficient to reduce any added utility from obeying the conversational maxims. A simple statement of the time takes the FTA off record, allowing me to draw my own conclusions from a statement of the time. Amy's preferences are likely to be such that the utilities work out as follows:

$U_{Amy}(\textit{It's five minutes to four}) > U_{Amy}(\textit{Robin! Go to your meeting!})$.

Notice that the complexity and length of the utterances are comparable, so the deciding factor here is likely to be face.

Of course, there is always the risk that I won't apply the information to the decision problem and make the wrong choice. Instead of concluding that I should go, I might simply note that time and continue with whatever I was doing, blithely missing an important meeting. Amy has assumed that I will be cooperative and draw the correct inference, but—as with the stag hunt game—there is some risk that I will simply opt for the lower payoff.

My grasp of the game may be defective or at least different from Amy's (see chapter 6). Thus, I might not realize what hinges on the time. It's quite likely that I have simply forgotten about the meeting or its being scheduled for today. If Amy has reason to suspect that I won't get the uptake, she might opt to go on record with the face-threatening act by adding some redressive act:

(42) Sorry to disturb you, but you have just enough time to get to the meeting; I know you wanted to be there.

The politeness formulas redress my positive face (by validating my desire to go to the meeting) and my negative face (by apologizing for the imposition on my freedom to act). Thus, the utilities must work out as follows:

$U_{Amy}(\textit{It's five minutes to four}) > U_{Amy}(\textit{Sorry to disturb you, but}\ldots)$

$> U_{Amy}(\textit{Robin! Go to your meeting!})$.

That is, all else being equal, Amy would prefer to do the FTA off record; in the absence of that, she will perform the FTA openly but will offer redress in the form of either positive or negative politeness or some combination of the two. Brown and Levinson's equation for the weight of a face-threatening act can be modified into a quick computation of the cost of a face-threatening act to utility:

(43) Cost(act) = Weight(act) − Redress(act).

We might, then, compute the utility of an utterance for a speaker as a function of the success of the utterance plus the degree to which it obeys the conversational maxims less the cost as computed in (43).

We should consider a computation for the utilities when Amy simply intends to tell me the time and not perform an FTA off record. Recall that in this case she can either simply announce the time or announce the time and explicitly cancel the implicature that I need to go to my meeting.

Working out the preferences here can be a bit sticky. It might be a matter of indifference to Amy whether I connect her announcement to my decision problem and make the inference that I should go to my meeting. In fact, though, given the choices at hand, we should assume that Amy actively does not want me to make the inference, since doing so would involve my inference that she had threatened my face. Thus, she entertains the possibility of canceling the implicature completely.

This brings up the point that politeness, including the possibility of doing an FTA off record, is highly dynamic. Depending on strategic circumstances, a person might be indifferent to performing an FTA off record, or in a slightly different circumstance, she might have an absolute horror of performing the face-threatening act even if it is off record. Suppose that Amy really wants to avoid implicating that I should go to the meeting; she might prefer the longer expression that cancels the FTA to the risk that I might take offense.

Figure 8.9 shows an example of how to work out the utilities for the game. For convenience, I've decorated the tree with annotations concerning how the face-threatening act was taken, whether there was an implicature, and so on. The highest utility is associated with the case where Amy manages to successfully do the face-threatening act off record. Notice that the case where she intends to implicate that I should go, but I fail to go, gets the lowest utility; information was transmitted, but the intended communication failed. Doing a bald on-record FTA gets only slightly more utility than failure; the consequences of threatening my face outright are too dire, although the act does have the intended effect.

If Amy just wants to tell me the time, she risks my making the inference and taking it as an FTA, albeit indirect. If she wants to avoid my indignation, she might want to explicitly cancel the implicature.

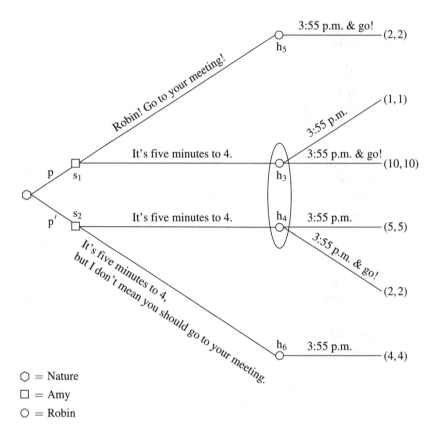

Figure 8.9
Implicature and Politeness

Given the way I've assigned utilities in figure 8.9, Amy's best bet, if she wants me to go to the meeting, is to do the FTA off record by telling me the time. If she wants to tell me the time without any risk of a politeness contretemps, then she should tell me the time and cancel the implicature that she thinks I should go to the meeting.

Of course, there are a number of ways to assign utility depending on how the participants view the hazards and likelihoods of various face-threatening acts. The following is a schematic for working out utilities:

Maxims(act) − Cost(act),

where Maxims(act) is the sum of the conversational maxims that the act obeys, and Cost(act) is given by its weight less the redress accorded by the speaker to the hearer. The weight, in turn, is based on Brown and Levinson's equation, which takes into account the power relationships between the speaker and the hearer, their social distance, and the relative gravity of the act in the culture.

I would argue, following Brown and Levinson, that face-threatening acts are an important ingredient in understanding conversational implicature. There is, though, a greater contribution that the study of face can make. The focus so far has been on threats to face; our social life, however, involves not only threats to face but also the active construction of face. Recall that chapter 7 presented an example of irony, in which one of my colleagues said, in response to the department chair's calling a faculty meeting,

(44) That sounds like fun!

Of course, I knew that he meant the exact opposite of what he literally said.

The original discussion used the example to show that mutual information, in the form of sympathy, was crucial in working out irony. This is surely true; I could not have understood my colleague's meaning unless I understood that he does not, in general, look forward to faculty meetings. I could not have constructed a coherent game tree—one compatible with my knowledge of the world and past experience—if I had considered only the literal content of his utterance. In order to work out the game, I needed to consider more than the literal content of the utterance.

In cases of irony, the implied content of an utterance contradicts the literal content. Why would anyone choose to take such a risky route to signal a simple meaning? There must be some gain in utility that makes the risk of miscommunication worth the trouble. Of course, my colleague knows me well enough to know that I won't be taken in by his utterance. That is exactly the point of his utterance. By saying the opposite of what he meant, he constructed a backchannel message that included me in a small cabal of fellow sufferers. In this way, he reinforced a social network that can be called upon in the future. If I get his ironic statement, then we both share in some added utility from the construction of this social network. This game is shown in figure 8.10 with some somewhat facetious labeling. The idea is essentially the same as in the original analysis with the added twist that there is extra utility for successfully using irony to communicate an intended meaning, precisely because the use of irony builds a social link and adds to the speaker's face.

Here we arrive at the crucial point: in using language we are constantly constructing face. We do so not just by face-threatening acts but

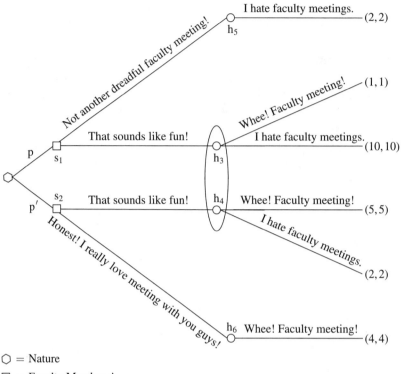

Figure 8.10
Irony and Face Building

by face-building performances. Language allows us to construct and maintain networks of mutual interest, reciprocity, and social alliance. The construction of these systems involves face, both to make calls on individuals and to bank face against the future, when such calls might be needed.

We have returned, then, to the discussion of the evolution of cooperation, a topic broached in chapter 4. The standard analysis of cooperation in terms of evolutionary game theory (see the discussion of the hawk-dove game) reduces to Prisoner's Dilemma. This approach is too weak to account for real cooperation. The construction and strategic use of face seems like a plausible method of attacking the problem. In particular, we might hypothesize that face developed from dominance hierarchies in primate groups. When we combine this viewpoint with coordination games—particularly the use of focal points (see chapter 9)—we can see how public coordination might evolve into a full-fledged theory of mind.

On Game-Theoretic Analysis

It is worthwhile pausing here to summarize the ingredients of game theoretic-analysis. The method itself is quite straightforward. The first step is to consider the problem in terms of the choices that the speaker and the hearer make. These choices are often made available by a combination of the grammar and the local context available to both the speaker and the hearer.

For example, with discourse pronouns, the local model—the discourse model—makes a certain number of discourse entities available to the speaker and the hearer. This information is available to both and thus should be represented in the game tree. Equally, both the speaker and the hearer know that, all else being equal, the speaker could use either a pronoun or a description to refer to the object in the discourse model.

The game can be solved by working out the preferences of the speaker and the hearer and mapping these onto utilities. This is the crucial step in any game-theoretic analysis. In the absence of a constrained theory of preferences, we are free to assign utilities any way we like. The analysis then risks becoming a just-so story, entertaining to develop but ultimately unconvincing.

The example of conversational implicature shows the importance of developing a theory of preferences for the problem at hand. As noted, preferences can be ranked in terms of the following:

- Truthfulness, orderliness, and efficiency of the utterance (the conversational maxims)
- Contribution of the utterance to solving a decision problem
- Threat to face of the utterance
- Potentially, the construction of face

We can evaluate utterances to the degree that they obey these requirements. The result is a full game tree that can be solved by the players.

The advantage of this analysis is that it embeds communication in a theory of rational decision making. The theory employs grammar but is not embedded in grammar. I would argue that a full theory of meaning can only be developed in light of the interaction of rational agents in a context, that is, a theory of rational language use. We might hypothesize that the semantic content of an utterance is a very sparse representation of meaning. The full meaning of an utterance is, in fact, the result of strategic negotiation between speakers and hearers. Chapter 9 examines a theory of strategic appropriateness.

A final word about probabilities would be useful here. I have not said much about the assignment of probabilities to the various information states, although the question is important. Probabilities can be used to tune the expected utilities, causing solutions to "pop out" depending on the agents' estimation of the likelihood of one or the other information state.

As a rule of thumb, I assume that the initial information states, the ones where the speaker's intended meaning is fixed, are equiprobable. Fixing the probabilities this way means that the utilities must be fixed in order to solve for the optimal strategy profile. Once the utilities are fixed, we are not allowed to change them; the pop-out solution is then a matter of the probabilities. By keeping the utilities fixed, we are forced to find a principled theory of preferences. Only then can we counter the charge that the models are nothing more than just-so stories.

Further Reading

The analysis of discourse pronouns in this chapter is basically that of Clark and Parikh (2007). Mayol (2009) extends the basic theory to a language with an extra resource, null subjects. Some of this work is reported in Mayol and Clark (2010).

There are, of course, many available analyses of discourse anaphora. Centering theory, for example, has been very influential and has been an influence on the game account. See Grosz, Joshi, and Weinstein (1983; 1986) for the early development of the theory. Walker and Prince (1996) give a very approachable discussion of the theory. The papers collected in Walker, Joshi, and Prince (1998) are particularly useful. Beaver (2004) reformulates centering theory in terms of optimality theory. Like game theory, optimality theory is a method of optimizing certain functions; game theory's focus, however, is on strategic interaction, taking into account the potential actions of two (or more) agents, whereas optimality theory lacks this strategic character.

Centering theory has been critiqued by relevance theorists like Breheny (2002), who provides a very useful resource with a number of interesting examples.

Grice's influential 1975 paper can also be found in the collection Grice (1989). A great deal of work after Grice focused on simplifying the maxims and eliminating redundancies in their statement. See Levinson (1983; 2000) for a discussion. Horn (2001) provides a classic discussion of the maxims. I have not considered these reformulations here, staying within Grice's formulation because of its accessibility, but a full working out of the utilities would be well served by a more careful consideration of the later work.

The classic text on politeness is Brown and Levinson (1987). Their work builds on the work of Erving Goffman; see Goffman (1959; 1967). Goffman (1967) contains a useful discussion of the notion of face. In general, his work proves important for the whole project of working out the construction and maintenance of face.

Van Rooy (2003) approaches politeness from the perspective of evolutionary game theory. Polite forms tend to be more complex and thus more costly; see Zahavi's (1975) handicap principle, which suggests that reliable signals must be costly to the signaler. Since these signals cost the signaler something that could not be afforded by an individual with less of a particular trait, evolution will tend to lead to reliable signals.

9 The Social Ecology of Meaning

Chapter 7 presented games of partial information as a general model of communication in natural language. This class of games makes possible a very fine analysis of ambiguity and disambiguation. The context can affect the subjective estimate of probabilities, allowing one or another ambiguous form to "pop out." Downtown in the financial district, the word *bank* will most likely be interpreted as a financial institution. Out in the country near a river, *bank* is likely to be interpreted as the side of a river.

There is a lot more to be said about lexical items and lexical meanings beyond accounting for their ambiguity. In this chapter, I explore some other aspects of lexical items.

There is a half-truth that I told in chapter 5 for which I now want to make amends. In that chapter, Abélard, the falsifier, and Eloïse, the verifier, played a game of picking and choosing objects from a world model in order to verify or falsify a sentence. I pretended at the time that both the verifier and falsifier had perfect access to the set of elements that denoted by a word. For example, if the sentence was

(1) Some dog barks.

then both players had perfect access to the things in the world model that count as dogs. In other words, if something is a dog, then the players know it is a dog, and if the players know something is a dog, then it is, in fact, a dog. In general, this can be simulated by listing the things that a word denotes. Of course, in a full semantics for natural language, words can denote all sorts of peculiar things like functions from sets to sets, and so on. For the moment, though, I want to focus on simple concrete nouns.

Let's take *dog* as an example. We can simulate knowledge of what *dog* means by simply listing the things in the world that count as dogs:

(2) DOG = {Sami, Faye, Ginsberg, Rover, MacDuff, Sandy, Pip, Fido, Adorno, Flip, Blue, Tucker, Apple Sauce, Two Dot, ...}.

Of course, I can't list all the dogs in the world, even if I could in principle pick them out accurately. This points out a weakness in this kind of treatment of word meanings.

The idea of listing the extension of a word—the set of things that the word denotes—is a mathematical trick; the idea is to simulate the traditional notion of an *Aristotelian definition*. An Aristotelian definition would give some set of necessary and sufficient conditions that things must have in order to count as instances of whatever the word denotes. That is, anything that satisfies the conditions in the definition would count as an instance of the word and nothing else would count as in instance of the word. An Aristotelian definition should cover all and only the things denoted by the word.

It's clear, though, that this isn't how words work. I, for one, know the Aristotelian definitions for almost no words. I failed to give a useful definition for *tiger* (concluding, in fact, that tigerness was a primitive essential property) and admitted that I know almost nothing about trees and plants. I know that dandelions are flowering plants, but I wouldn't be able to pick one out in a garden with any degree of accuracy.

Things get worse when we look farther afield. Consider a word like *traffic*, as in

(3) The traffic was light on the Atlantic City Expressway today.

What does the word *traffic* mean here? Presumably, some density of number of cars on a road. How many cars count as traffic? Two, a dozen, more?

Well, that all depends on the road. If I'm driving around central Baja California, then two or three cars can count as traffic. There aren't many cars on the road down there. But what about the New Jersey Turnpike? If there are only two or three cars around me on the New Jersey Turnpike, then that's not any traffic at all. The right way to think about the meanings of words and phrases is not in terms of necessary and sufficient conditions but rather *appropriateness of usage*. Does the word signal the intended meaning to the hearer in the context? Thus, working out appropriate usage involves games and connects us with the question of how we can communicate in the face of the ambiguity regularly found in natural language.

Games and Prototypes

The analysis so far has been based on the fiction that words and phrases straightforwardly pick out things like sets. So, for example, when I use *pen* in the sense of writing instrument, I am denoting the set of pens (for writing); when I use *dog*, I am denoting the set of dogs; when I use *old world monkey*, I am denoting the set of old world monkeys, and so on.

This treatment of word meaning is a useful fiction to the degree that it gets the study of meaning off the ground. It allows us to incorporate a comfortable idea about definitions into the formal theory of meaning using a very simple mathematical object: a set. The intuition is that if I, as a competent speaker of English, have mastered the meaning of a word, then I should be able to separate examples of the word from things that are not instances of the word. So sit me down, say the word *tiger*, and then parade objects before me. I should say "yes, tiger" whenever a tiger passes in front of me and "no, not tiger" when a non-tiger is presented. I seem to have somehow mastered a set of necessary and sufficient conditions for tigerhood that allows me to group all and only tigers under the word *tiger*. In other words, sets (or characteristic functions) provide a formal means of realizing Aristotelian definitions.

Except, of course, that the world doesn't really work that way. We shouldn't be satisfied with a comfortable definition if all it does is gloss over interesting data and significant generalizations. In fact, as I argued in part I, speakers often don't know which objects fall under a word. We all rely on other people to fix the meanings for us, and each of us is often uncertain about how to apply words to things. Meaning is an emergent property of social systems. How this happens is a fundamental explanatory problem for linguistics.

Sit me down, now, along with another native speaker, say the word *bald*, and then parade things before us. Let us write our decisions down on scraps of paper and see how often we agree and disagree. Suppose men pass in front of us. We'll often agree on bald and nonbald men, but there will be a significant number of men who are somewhere between bald and not bald on which we might disagree. The boundaries of baldness are vague, with a more or less vast indeterminate area between the clearly bald and the clearly not bald.

Now start passing other objects in front of us. Is an apple bald? a coconut? a basketball? How about a chair? Sometimes it seems that applying *bald* to something is inappropriate. Sometimes an interpretation must be constructed out of the parts. Suppose, for example, you come to my house, and I say, gesturing toward a chair,

(4) The bald chair is over there.

You would almost certainly arrive at an interpretation for (4). Perhaps the chair is a special one for seating bald people; you might conclude that I had implicated you were bald. You might, on the other hand, wonder if I didn't have some hairy chairs—chairs covered in hair—and if this was then the only bald chair in my house. Some people I've asked have interpreted *bald chair* as one that has threadbare spots, a perfectly sensible interpretation.

We construct interpretations when called upon to do so, and these constructions are cooperative. Our daily speech is full of loose talk; the speaker attempts to find an expression that will signal her intended content to the listener. The listener cooperates by trying to work out what the speaker must have meant by saying thus and so. Part of our job in understanding how language works is working out how speakers and hearers cooperatively construct interpretations.

Van Deemter (2010) argued persuasively that vagueness is the norm for language. Supposedly precise scientific terms, like *species*, have vague boundaries. If we suppose that members of the same species can interbreed, for example, then we can construct a chain such that members of each adjacent link in the chain can interbreed but the ends of the chain cannot. Where does the species boundary lie in this case?

Even constructed terms like *meter* are inherently vague, as van Deemter showed. They rely on our ability to measure various quantities. Originally, a meter was defined relative to the meridian of the Earth that passed through the North and South poles via the Panthéon in Paris. Now, it's defined relative to the distance light travels in a vacuum in 1/299,792,458th of a second. However precise we try to make the meter, these measurements are subject to errors and approximations.

Vagueness is the norm in language. In science, mathematics, and philosophy, we try to eliminate vagueness by stipulating ever more precise constructions. These constructions, though, are arrived at by the social practice of science, mathematics, and philosophy. In order to understand how this process of construction works, we must come to grips with vagueness. It seems odd to start with the assumption that linguistic elements have precise Aristotelian definitions, modeled by sets, and then try to impose vagueness; the world moves in the opposite direction.

The philosopher J. L. Austin got at the intuition very well:

> Suppose that we confront "France is hexagonal" with the facts, in this case, I suppose, with France, is it true or false? Well, if you like, up to a point; of course I can see what you mean by saying that it is true for certain intents and purposes. It is good enough for a top-ranking general, perhaps, but not for a geographer. "Naturally it is pretty rough," we should say, "and pretty good as a pretty rough statement." But then someone says: "But is it true or is it false? I don't mind whether it is rough or not; of course it's rough, but it has to be true or false—it's a statement, isn't it?" How can one answer this question, whether it is true or false that France is hexagonal? It is just rough, and that is the right and final answer to the question of the relation of "France is hexagonal" to France. It is a rough description; it is not a true or false one. (Austin 1975, 143)

Truth is only one aspect of meaning, a single metric from a larger array that we can use to judge and interpret utterances. It is not even clear whether truth is the principal metric for judging meaning; perhaps a better metric would be appropriateness. Does the word fit what I'm trying to get across to my interlocutor as we collaborate in constructing meaning from utterances? When we talk, we are engaged in coordination games where we construct and elaborate meaning; while truth is certainly a component of this project, is it the only one or even the best one? It is these intuitions that a real theory of meaning must grapple with.

Prototype theory is one compelling alternative to the Aristotelian approach to definitions. The theory is organized around two guiding principles: *cognitive economy* and *perceived world structure*.

(5) **Cognitive Economy**
   The task of category systems is to provide maximum information with the least cognitive effort.

In essence, a category is an attempt to extract as much information from the environment as possible. In other words, given a few features of the object, I want to be able to extrapolate as much information as possible about that object.

(6) **Perceived World Structure**
   The perceived world comes as structured information rather than as arbitrary or unpredictable attributes.

This principle implies that there should be a correlational structure between perceived attributes.

It is perhaps easiest to consider an example of a category, like "bird." Certain features are stereotypically associated with birds; they have wings, feathers, and beaks; they have a particular body shape; they have peculiar clawed feet; they fly; they lay eggs. When asked to list the properties of birds, subjects reliably list properties like these. These features have high cue validity, that is, their presence reliably indicates the presence of a bird. If I note that something has a beak, then the probability that it is a bird is quite high. I can, in fact, use this feature to predict not only the presence of the other features but that the object is actually a bird.

When I think of a bird, I'm likely to think of a prototypical instance of a bird, that is, I'm likely to come up with sparrows and robins and canaries, not emus or pelicans or penguins. Real-world instances of birds will distribute around the prototypical type, with most birds clustered around the prototype but with a certain number of outliers, birds that are real birds but not like prototypical birds.

Notice that absence of some of these features does not mean that the creature in question is not a bird. Thus, pelicans have a peculiar body shape. Chickens are birds, but they are built strangely and are bad at flying. Penguins and ostriches are also shaped oddly and don't fly at all. Finally, presence of a feature does not necessarily predict inclusion in the concept class; a platypus has a beak and lays eggs but is not a bird.[1]

One might maintain that prototypes define fuzzy categories, that is, membership in a category—the extensional counterpart of a concept—is measured by distance from the prototype. Categories would have fuzzy boundaries, and membership in a category would be probabilistic. Thus, a pelican would be, say, 87 percent a bird because of its peculiar body morphology.

Furthermore, as noted in chapter 1 with respect to Claude, the albino, bipedal, vegetarian tiger, people behave like essentialists with respect to natural kinds like tigers. They suppose that the defining property of tigers is something like tiger DNA. Of course, you only knew Claude was a tiger because I told you he was. If I sat you down to categorize tigers and non-tigers, you might say "no, not tiger" when Claude passed before you unless you knew his exotic history.

The prototype approach to concepts is not quite right. It resurrects the internal mental language of Aristotelian properties in the guise of probabilities. There's little reason to suppose that a pelican is only partly a bird; it's just not a very good example of a bird. In addition, as demonstrated by Armstrong, Gleitman, and Gleitman (1983), concepts that we wouldn't want to call fuzzy at all have prototypes. For example, "even number" and "plane geometry figure" show prototype effects, although there is no sense in which 4 is a better even number than 358, or a square is a better plane geometry figure than an ellipse, although subjects might rate 4 as a better even number than 358 or rate squares as better plane geometry figures than ellipses.

Armstrong, Gleitman, and Gleitman included "female" among their well-defined categoriesm, good examples being aunts and ballerinas, and poor examples being widows and waitresses. Of course, one has to wonder how well-defined "female" is as a concept. Consider the case of the South African runner Caster Semenya. She was raised as a woman but after doubts were raised about her gender, testing revealed that she was, in fact, intersex. There are a number of causes for unconventional gender expression. Apparently, the category of "well-defined category" is itself fuzzy. As I suggested before, precise categories are socially constructed from ordinary language.

However, the prototype approach is insightful about how we *use* categories. Putnam (1975) argued that we use prototypes as identification procedures. Armstrong, Gleitman, and Gleitman had the same intuition. Putnam argued for a classic conception of categories, that is, meaning as regulated by the division of linguistic labor (see chapter 3). Social practice regulates meaning, so we rely on experts to tell us who the real tigers are, if the question arises.

The usual treatment of prototypes often talks in terms of features in an internal mental calculus, a variant of the Mentalese hypothesis that I argued against in part I. We need to take into account our ability to coordinate our linguistic behavior socially. The next section reconstructs prototype theory in terms of coordination games. Prototypes correspond to highly salient features of the world around which we can coordinate our symbolic behavior. In that sense, coordination games can translate apparently private concepts into public behavior.

Metrics, Central Tendencies, and Focal Points

There can be little doubt that we extract from our experience concepts that we use to organize and categorize our day-to-day affairs. The resulting concepts support our use of words, but they are not identical to them. Concepts can cross-classify words. Words are grouped into semantic relations—inclusion, synonymy, antonymy—that we can organize using concepts. I suggest here that concepts and word meanings in general involve games that use a central tendency and a metric to organize objects under concepts.

By *central tendency* I mean something like a prototype, where a prototype represents a best example of a category. We need not suppose that the central tendency corresponds to an actual object, just as the average of a list of numbers need not be a number on the list. Indeed, it is easy to find examples in natural language that do not—and are not intended to—pick out actual things:

(7) The average American woman has 2.1 children while the average woman from Niger has 7.29 children.

Certainly one would be hard put to find a woman anywhere who has had 2.1 children or 7.29 children. While averages are not the same as prototypes, the point stands; prototypical concepts need not correspond to actual objects in the world.

The combination of a central tendency plus a metric is intended to explicate the idea of appropriateness of use. A term appropriately applies to an object when that object is taken to be sufficiently close to the central tendency associated with the term according to the metric. In order to coordinate our behavior around an object, we need to agree that the object appropriately falls within the domain of the term; reference has a social basis.

Suppose I say,

(8) Look at that bald man.

You look and see a variety of men of varying degrees of baldness. Some of them may have a lot of hair, while others have less. Perhaps one has just a few strands of graying hair awkwardly arranged in a comb-over. Now, the canonical example of a bald person might be a man with absolutely no hair at all. In the scene you are looking at, nothing quite fits. Nevertheless, the guy with the comb-over comes closest to what you might take as bald. Measuring the distance from the comb-over guy to unquestionable baldness reveals that, indeed, Mr. Comb-Over is someone you will take as bald.

Another example might be germane. My cocker spaniel Sami dislikes the cold, so we tend to let his hair grow in the winter until he is quite hirsute. It has happened that my wife said to me,

(9) The woolly mammoth needs a walk.

I know exactly who she means; Sami, in full fur, does resemble a miniature woolly mammoth; all that's missing is the trunk and tusks. Of course, Sami isn't really a woolly mammoth—what she said was false—but there is no doubt as to what she meant. Sami has some of the *salient* properties we would associate with a woolly mammoth—big ears, a portly profile, and an abundance of hair. Except that he's a 35-pound cocker spaniel and not a woolly mammoth at all. But, of course, my wife never lies!

Observing my environment, I find one party that has the salient features of a woolly mammoth; applying my metric, I conclude that he is the closest thing we have at my house to a woolly mammoth, and I'll agree that Sami can appropriately count as a woolly mammoth for present purposes. Time to get the leash and my coat. Is this an odd use of language? Not at all; it seems quite normal to me, a bit of the quotidian poetry we all play with and thus as likely a datum for semantico-pragmatic theory as anything else.

It's worthwhile here to recall some classic studies by Posner and Keele (1968; 1970). Posner and Keele generated a pattern of random dots that served as the prototype of a category. Using this pattern, they generated a large number of new random dot patterns, which they called distortions, by moving each point in the original in a random direction, sometimes moving the dots a small amount and sometimes moving them a lot. Of course, all the distortions were similar to each other by virtue of being generated from the same underlying pattern.

Subjects who had viewed only the distortions of the original pattern were able to learn the category, that is, they could correctly distinguish dot patterns that were distortions from other dot patterns. Indeed, although the original prototype dot pattern was not included in the original set of images presented to subjects as part of learning the new category, the subjects were able to identify it as a member of the category during testing. What's more, items that were made by small distortions were learned better than items that were made by large distortions. Subjects who were tested with new items made by small distortions were also more accurate at correctly categorizing them. In general, the smaller the distortion, the more typical the item was, and the more accurately it was classified.

Of course, one can only conclude so much from concepts based on random dot patterns, learned under artificial conditions in a laboratory. But consider another classic study by Labov (1973). A number of line drawings were prepared of household receptacles of different shapes. These drawings were then shown to subjects who were asked to name the objects depicted in the drawings. An item with a circular horizontal cross-sectional area tapering toward the bottom, whose maximum width was equal to its depth, and which was provided with a handle, was unanimously judged to be a cup. As the ration of width to depth increased, more and more subjects called the object a bowl. But there was no clear dividing line between "cup" and "bowl"; subjects differed as to the point at which they would switch from one term to the other. Equally, removing the handle lowered the likelihood that an item would be called a cup, but the effect was not clear-cut. If subjects were asked to imagine the receptacle as full of coffee, they were more likely to call it a cup; if they imagined it full of mashed potatoes, they were more likely to call it a bowl. So various aspects of the environment influenced judgments.

All of this is consistent with the idea that there are central tendencies around which our concepts are organized and metrics that help us apply these concepts to real-world objects and events. As an illustration of the use of metrics, consider a verb like *crawl*, which indicates a particular type of locomotion, movement on the hands and knees dragging the body close to the ground. Nevertheless,

(10) a. The baby crawled on the floor.
b. The caterpillar crawled along the leaf.
c. The car crawled along the mountain road.
d. The train crawled into the station.
e. The evening crawled by.

All the sentences in (10) seem to be fair uses of *crawl*, although the type of locomotion is different in each up to the last case, which isn't, strictly speaking, locomotion at all. They illustrate the polysemy of *crawl*; note that the various polysemous uses of *crawl* are clearly semantically related.

All these cases have a similar structure. In the case of *crawl*, there is a central or core type of crawling, perhaps to move slowly on the hands and knees with the abdomen or body close to the ground. The properties associated with this kind of crawling can be simply enumerated; the extensions of *crawl* would be organized around this central focal point, perhaps by removing one or more of the properties.

The notion of prototype—or central tendency—is related to Schelling's (1960) idea of tacit coordination. Schelling observed that there is a form of implicit bargaining under conditions of limited or no communication. Such bargaining would work by organizing it around "obvious" points that each participant could assume were found generally obvious by all the other participants. He called these points focal points, but they have come to be known as Schelling points. Recall that focal points were part of the discussion of common knowledge in chapter 6.

Suppose that a couple becomes separated in a department store without having first arrived at a plan as to what to do in such an eventuality. They will, of course, try to think of obvious points in the department

The Social Ecology of Meaning 293

store where the other is likely to go. That is, they are engaged in tacit coordination in an attempt to rejoin each other. Notice that both participants need to coordinate their choices; one person's simply picking his favorite spot is useless unless he has reason to suppose that the other person will also select that spot as a possibility.

Schelling was quite clear on what a focal point is:

> These problems are artificial, but they illustrate the point. People *can* often concert their intentions or expectations with others if each knows that the other is trying to do the same. Most situations—perhaps every situation for people who are practiced at this kind of game—provide some clue for coordinating behavior, some focal point for each person's expectation of what the other expects him to expect to be expected to do. Finding the key, or rather finding *a* key—any key that is mutually recognized as the key becomes *the* key—may depend on analogy, precedent, accidental arrangement, symmetry, aesthetic or geometric configuration, casuistic reasoning, and who the parties are and what they know about each other. Whimsy may send the man and his wife to the "lost and found"; or logic may lead each to reflect and to expect the other to reflect on where they would have agreed to meet if they had had a prior agreement to cover the contingency. It is not being asserted that they will always find an answer to the question; but the chances of their doing so are ever so much greater than the bare logic of abstract random probabilities would ever suggest. (Schelling 1960, 57)

Schelling gave a number of examples of what he meant; let's consider a few examples from Schelling's list. In each of the following puzzles, you are to coordinate with another person, but you have no way of communicating with that person to negotiate an answer. This is, in other words, one-shot bargaining:

1. Name "heads" or "tails." If you name the same element as your partner, you will both win a prize.
2. Circle one of the numbers listed in the line below. You and your partner will both win a prize if you circle the same number.

7   100   13   261   99   555

3. You are to meet someone in New York City, but have not been instructed where to meet; you have no prior understanding with the person about where to meet and you cannot communicate with each other. Where do you go?
4. As in the previous puzzle with the added twist that although you know the date of the meeting you don't know the time. What time do you show up?

Schelling did an informal poll to see what people would answer. For puzzle 1, thirty-six people chose "heads" while only six chose "tails." For puzzle 2, the first three numbers (7, 100, 13) received a total of thirty-seven votes out of forty-one; 7 led 100 by a slight margin, with 13 in third place. In puzzle 3, an absolute majority selected the information booth at Grand Central Station. Finally, in puzzle 4, an absolute majority selected noon. Notice that these answers are not based solely on logic, as Schelling observed; rather, we have biological and social equipment that biases our answers.

Schelling went on to consider focal points in coordination games where the interests of the players did not converge, that is, one player got a higher payoff for one answer than the other player did (and vice versa for some other answer). He again found some evidence in his informal polling that respondents attempted to coordinate their answers even under these conditions—after all, a mediocre payoff is better than no payoff.

Schelling's informal experiment was replicated with tight controls by Mehta, Starmer, and Sugden (1994). They reasoned, following Lewis's (1969) work on conventions, that people might choose on the basis of *primary salience*, that is, for some reason, an option comes into the subject's mind—for example, the number 7 in puzzle 2 might be considered a lucky number. The subjects simply pick what is, for them, the most salient option without any intention of coordination per se; people would use primary salience to somehow muddle through in coordination problems.

Another possibility is that subjects try to coordinate via *secondary salience*, each player choosing according to what she believes would be of primary salience to her partner. Thus, her choice has secondary salience for her; she simply imagines what is most salient for her partner.

Finally, there is the possibility that subjects use something like Schelling's concept of salience. Mehta and colleagues (1994, 661) summarize it as follows:

> When someone is playing a pure coordination game, she will look for a *rule of selection* which, if followed by both players, would tend to produce successful coordination. A rule of selection... is salient to the extent that it "suggests itself" or seems obvious or natural to people *who are looking for ways of solving coordination problems*. We shall call this conception of salience *Schelling salience*.

Notice that Schelling salience is different from secondary salience. In the latter, a person tries to imagine what her partner would pick. In Schelling salience, a person tries to imagine a general coordinating rule that any reasonable person would also notice.

The experimenters divided subjects into two pools. The subjects in group C (for *coordinating*) were instructed to make their selection with the objective of giving the same answer as an unknown person with whom each had been paired. The instructions explained that a pool of money would be divided between the members of the group, each subject's payment being proportional to the number of points he had won by successfully coordinating. Thus, subjects in group C had a motivation for coordinating their answers.

The subjects in group P (for *picking*) were simply told that they were answering a survey. There were no monetary consequences for their choices. The subjects in the P group had no reason to attempt to coordinate their responses.

After answering the surveys, the results were randomly paired within groups to see how often the subjects coordinated. That is, the responses by subjects in group C were randomly paired with each other, and the responses by subjects in group P were randomly paired with each other. Mehta, Starmer, and Sugden found a highly significant effect for coordination. The mere fact of being told to coordinate was sufficient to increase subjects' coordination.

Although they did not attempt to distinguish between secondary salience and Schelling salience, the experimenters found some evidence that Schelling salience was, in fact, the appropriate concept. For example, when subjects picked a positive integer with no attempt to coordinate, 7 was chosen 11.4 percent of the time, followed by 2 at 10.2 percent of the time. When subjects sought to coordinate, 1 was chosen 40 percent of the time, even though it was not primarily salient for most people (1 was chosen only 4.5 percent of the time in group P). Clearly, more experiments need to be done to work out the differences between secondary salience and Schelling salience, as well as to test the effects in games where the interests of the players conflict with each other.

Nevertheless, it is clear that coordination has a significant effect on how people make their choices. When called upon to do so, we have an uncanny ability to coordinate our behavior in the absence of any prior agreement.

The main hypothesis is that prototypes are points in the semantic space that are focal in the sense that speakers will coordinate their communicative behavior around them. This makes sense if one supposes that there can be no communication without tacit bargaining. When I say something, I make a strategic choice based on my assessment of the likelihood that you will coordinate with me. I might use a particular expression—for example, "the dance troupe performing at the Prance this weekend"—anticipating that what I have in mind is what you will pick. In lexical choice, I will choose my words in such a way as to maximize the likelihood that we will coordinate around my intended meaning. There can be no communication without coordination, and coordination is facilitated by finding focal points.

In trying to coordinate, a good rule of thumb might be to pick the prototype, particularly if prototypes are learned from past instances of linguistic coordination. There is a difference in response between free choice and coordinated choice, as is evident in Schelling's informal poll and in the more tightly controlled study by Mehta et al. For example, when asked to name a flower, subjects in the P group selected the following as top responses:

Response	Proportion
Rose	35.2
Daffodil	13.6
Daisy	10.2
Tulip	9.1

Roses are fairly prototypical flowers, as are daffodils, daisies, and tulips. In this condition, subjects are asked simply to name a flower; they aren't given a fixed set to choose from, so they are free to pick anything. I might pick a Venus fly-trap for fun. Free choices can be quirky, even if there is a tendency to stay close to prototypical examples because they tend to come to mind the most quickly.

Consider what happens in the following coordination task:

Response	Proportion
Rose	66.7
Daisy	13.3
Daffodil	6.7

Again, the subjects are not given a fixed set to choose from. Notice the huge preference for saying "rose." The lesson is clear; when coordinating with others, choose a prototypical example.

I should note that Mehta et al. also looked at cases where prototypes didn't seem to be the obvious explanation. For example, they gave subjects geometric patterns and asked them to group the elements, the C group being instructed that if subjects coordinated their response with an unknown partner, they would be rewarded. Subjects were also given circles with various lines drawn through them—sometimes straight lines with one curved line or curved lines with one straight line. They were asked to pick a line that would divide the circle. Again, the C group was told that subjects would be rewarded if they coordinated with an unknown partner. Both types of cases showed an effect for coordination. In these cases, it would be hard to relate focal points and prototypes; instead, subjects seemed to be using rules based on "odd man out" or gestalt principles for grouping objects in a scene.

Clearly, a great deal of work remains to be done to determine the relation between focal points, prototypes, and coordination. Researchers need to control for prototypicality in both the free choice and the coordinating conditions, and for the frequency effects of words used to label the choices. The difference between secondary salience and Schelling salience could be explored, perhaps by manipulating mutual knowledge between the partners—if my partner and I know something special, know that the other knows it, and so on, we might exploit that knowledge to win the prize.

I'm proposing here that the prototype concepts associated with words correspond to focal points in a semantic landscape, that is, they are points around which we would naturally expect other speakers to coordinate in making choices.

Now, let's consider the second ingredient in the game-theoretic account of lexical meaning, the metrics associated with the focal points. The idea that concepts have focal points is not new. Rosch (1975), discussed a similar idea, although she did not relate the cognitive reference points to focal points in the sense used here. What I mean by a focal point is a point in semantic space around which linguistic agents can coordinate their behavior. I give the following slogan:

Prototypes are conventionalized focal points.

As noted in chapter 4, conventions are an efficient way of maximizing utility from a coordination game. Focal points are one-off solutions for coordination problems; once they have been repeated, they become conventionalized and serve as prototypical examples.

Let's return to the examples of *crawl* in (10) and consider how metrics can interact with the focal interpretation of an expression. Suppose that *crawl* in (10a) is an instance of the focal interpretation of *crawl*; that is, the prototype sense of *crawl* is to move slowly on the hands and knees with the abdomen or body close to the ground. In our semantic experience we encounter crawling events that match these criteria; babies, drunks, people looking for their contact lenses all crawl about. Indeed, we can measure the distance of various events from the focal interpretation. Presumably, events of crawling on the hands and knees would be tightly clustered around the focal point, while sprinting events would be quite far away. The metric allows us to measure how far from the focal interpretation we are willing to go before we say that the event in question is not a type of crawling.

A precise way of working out this idea would be in terms of vectors in a semantic space. Suppose there is a set of dimensions along which objects could vary. The result would be a hyperspace, with similar objects clustering together in the hyperspace. Focal elements would be the centers of these clusters. This idea owes a great deal to the work of Gärdenfors (2000).

The metric can be loosened by dropping the requirement that the motion be on the hands and knees. Any event where there is slow movement with the body or abdomen close to the ground will count as crawling. Now the locomotion of reptiles and insects as in (10b), as well as all the instances in (10a), count as crawling. Notice that the metric on what counts as crawling is lossened by dropping a requirement from the original set of focal properties.

To further widen the set by loosening the metric, we might drop the requirement that the body be close to the ground. These would be cases of slow movement. Then examples like (10c) and (10d) would be included as crawling. Notice a subsidiary prototype notion here, namely, slowness. Trains and cars move faster than babies or bugs, but the slow movement is measured in terms of the prototypical, or focal, movement of such things. Thus, a train appears to crawl as long as its motion is slow relative to normal train velocities.

Finally, there is the motion of time itself, as in (10e). Here the extension seems to be less a loosening of a metric than a metaphorical extension of the sense of *crawl* exemplified in (10c) and (10d).

Labov's (1973) experiment provides another example of the interaction between focal points and metrics. As noted, he asked subjects to sort drawings of containers into bowls or cups. The drawings near the focal point of the category, which was apparently determined by the ratio of the width of the opening to the depth of the container, were easily sorted, and there was wide agreement among subjects. The boundaries between cups and bowls were far from clear, however. Asking subjects to imagine that the container held coffee caused more drawings to be counted as cups; in effect, it loosened the metric around the focal interpretation of

The Social Ecology of Meaning   299

"cup." Equally, the instruction to imagine that the container held mashed potatoes widened the metric around the focal interpretation of "bowl," with the result that more of the containers were counted as bowls.

In the past, I have coordinated my behavior with other linguistic agents around things like cups and bowls. The things that were called cups could be sorted using the ratio of width to depth, so that my past encounters have a particular focal center. This is equally true of my past experience with bowls. Having learned to coordinate my behavior around such things, I know that "cup" and "bowl" are language games based on containers grouped around these focal centers.

You have had similar experiences playing language games with "cup" and "bowl" and have extracted particular focal centers for these things based on your experience. Presumably, our focal centers for "cup" do not match perfectly—your games involved some particular ensemble of cups that is different from mine.

Suppose I ask you for a cup. You look about for an object that will do as a cup. This might involve loosening your metric until you find an object in the local environment that will serve. When you do, you bring it to me. When I look at it, I also have to evaluate it as a cup relative to my focal center. If the cup you brought falls close enough to my focal center for "cup" according to my metric, then I'm satisfied. Reference, here, is ultimately conditioned by social factors; this is an instance of Schelling's idea about tacit coordination.

Suppose we're cooking together, and I say,

(11) Bring me the small bowl of mashed potatoes please.

You look about for a small bowl of mashed potatoes. You see a large bowl, but that won't do. You notice, near it, what appears to be a relatively large cup full of mashed potatoes. You bring it to me. The fact that the "cup" contained mashed potatoes allowed you to adjust your metric and accept what you would normally call a cup as a bowl. Reference doesn't fail, and we are still able to coordinate our behavior.

Consider an example of how context can shift the focal interpretation of a term. With a term like *fish*, I think there are two focal points, depending on whether we're talking about salt water fish or fresh water fish. In a fresh water context, I readily think of fish like bass or trout. In a salt water context, I think of marlin or tuna. Notice that there can be more than one focal interpretation for a term. Presumably, the context in which the term is used will draw attention to one most prominent focal interpretation.

What about a complex expression like *pet fish*? Here, the focal interpretation shifts yet again, with the focus of attention now centered on small fish like goldfish or tetras. The addition of *pet* shifts the focal interpretation in a way that is reminiscent of the garden path sentence in chapter 7, example (30). In that case, the presence of *astronomer* shifted the focal interpretation of *star* away from celebrities and toward astronomical objects.

Let's return to example (3):

(12) The traffic was light on the Atlantic City Expressway today.

What does *traffic* mean here? We might measure traffic by the number of cars within some stretch of roadway, say, m many cars per k meters. My focal center for *traffic* can shift depending on what kind of road we're talking about. I expect there to be a lot of cars on the Atlantic City Expressway, so my focal center for that kind of road is one thing. If we're talking about a back road in Lancaster County, Pennsylvania, more than a couple of Amish buggies might constitute traffic.

The discussion of *traffic* calls to mind the Sorites paradox, the reasoning that proceeds from the following premises,

1,000,000 grains of sand is a heap.
Taking away one grain of sand from a heap leaves a heap.

to the conclusion that just one grain of sand constitutes a heap, and taking it away still leaves a heap.

We have no grand social convention about what constitutes traffic, just as we have no convention about where heaps of sand end and small piles of dirt begin. Instead, we have experience, regulated by coordinative behavior. Labov's experiment with cups and bowls is really just a disguised form of the Sorites paradox; there is no clear point at which cups become bowls. There is just the ability to coordinate using linguistic signs.

The problem with Mentalese is that it presupposes a set of mental predicates connected to daily linguistic experience only by the loose ligaments of innate behavior. Real-world language use is flexible; language is used to coordinate our mental lives in a process of constant negotiation.

The basic idea can be stated briefly. Terms are associated with focal interpretations and metrics. The focal interpretation locates the core sense (or senses) of the term. The focal interpretation will normally be associated with a metric which, given a context, will determine which objects in the local model will be taken as falling within the domain of the focal interpretation. This metric can be shifted by the local linguistic context, as in *pet fish*, or by other aspects of the local context, as in referring to Sami the cocker spaniel as a "woolly mammoth."

As a practical proposal, one might try adapting game-theoretic semantics discussed in chapter 5 to take into account the notion of appropriateness. The semantic system discussed there was defined in terms of games of picking from a model. Logically complex expressions—quantifiers and logical connectives, for example—were replaced with simpler expressions until nothing was left except logically inert (atomic) expressions that could be directly verified against a model. The truth of a sentence was computed via a zero-sum game of perfect information.

Instead of giving the Abelard and Eloïse, the two players of the game, access to the extensions of predicates, we might allow them access only to the focal concepts of terms and some metrics that could be used in the context. Given the model, they would then have to agree on whether the object or relation counted for purposes of the larger game. That is, they would have to play a coordination game to see if they could agree. If no accord were reached, play would be suspended; there would be no winner in the larger game. Success in the coordination game would, under this interpretation, be a presupposition for the larger game. Notice that the overall interpretation of the game would remain the same: Eloïse and Abélard are working out the truth conditions of a sentence. That is how games of truth conditions could be translated into games of appropriateness conditions.

In the game approach to focal interpretations of concepts (and words), these interpretations are not simply a private language, an internal Mentalese. They are points around which speakers and hearers coordinate their linguistic behavior, and they are therefore fundamentally social. Focal interpretations and metrics are socially negotiated, much in the way that the value of currencies is socially negotiated (see part I). Speakers and hearers habitually coordinate their behavior around these points, and it is this social coordination that gives terms their semantic values. The next section describes how this negotiation might be carried out.

Semantic Landscapes and Meaning Niches

Schelling's original use of focal points included instances of physical space; maps and physical locations provide a useful metaphor for thinking about lexical meanings. We can think of a model—either the entire model or a partial model corresponding to the local context (the discourse context plus, say, a subpart of physical space that might provide salient objects for discussion)—as a kind of landscape. This landscape would be full of objects of varying kinds, and it would be a hyperspace defined by the objects in the space.

Imagine highlighting those objects that share some cluster of properties. We might let the degree of luminance of an object correspond to the degree to which that object shared the properties in question, that is, luminance represents the metric on the focal category. Objects with all the properties would be brightly lit. These would correspond to objects clustered around the focal center of the concept. Objects that are not lit at all would be objects that fall outside the domain of the predicate.

Since luminance is continuous, we can imagine a category like "bald" as having subtle gradations in luminance, with genuinely bald men having high luminance (and thus clustered around the focal center) and hirsute people having low luminance or none at all. If we compare "cup" and "bowl," following Labov's experiment, we might illuminate "cup" with one color and "bowl" with another; as the two categories shade into each other, luminance and hue would change, with intermediate cases being indeterminate as to which category they belong to. Metrics like "contains coffee" or "contains mashed potatoes" would alter hue and luminance.

Given a landscape, we might ask whether there is an optimal set of focal concepts plus metrics that optimally covers the space. That is, is there a way to cover all the objects in the space in a way that has the least overlap between concepts? Normally, there will be overlap between the territories staked out by the various focal concepts, the overlaps being partly determined by the metrics used to determine the extent of application of the various concepts (and their associated terms).

It is clear that terms can come into direct conflict. A famous example is the competition between *sheep* and *mutton*. The folk history is that *sheep*, which is derived from the Old English term for the animal, came into competition with *mutton*, derived from the old *moton*. According to this story, the Norman invasion brought a French-speaking aristocracy to English shores. As the folk story would have it, the Norman overlords would call for *mouton!* and the Saxon servants would dutifully go slaughter a sheep and prepare its meat for their masters. Accordingly, *sheep* has come to mean the animal, while *mutton* refers to the meat from the animal.

In fact, there must have been a competition between *sheep* and *mutton* to occupy a niche in the semantic space. Some remnants of this battle may still be going on; as recently as 1988 we find examples like the following (according to the *Oxford English Dictionary* entry for *mutton*):

(13) Leonora had had a mutton killed in anticipation of a family celebration.

where *mutton* clearly denotes the animal, in this case one intended for meat.

One finds a number of doublets like *sheep/mutton* in English. There are pairs like *pig* and *pork* or *cow/cattle* and *beef*. The latter case is particularly interesting. Growing up in West Texas, where there are many cattle ranchers, I remember hearing *beef* used as a count noun for *cattle*. Thus, *cow* was for a female beef, *bull* was for a male beef, and *cattle* was a mass noun for the collectivity. Of course, *beef* also denoted the meat of the animal as well, a clear case of polysemy. An example of the cattleman's usage can be found in Cormac McCarthy's *No Country for Old Men*:

(14) They put that thing between the beef's eyes and pull the trigger and down she goes.

We can imagine that initially the focal interpretation of *sheep* and the focal interpretation of *mutton* exactly coincided. They both denoted the same thing: the animal and its meat. Every time a speaker wanted to talk about the animal, say, she would have to choose between *sheep* and *mutton*. Every time she chose one, that would be a case where her choice was expressed—the frequency of *sheep* went up around that meaning—but the other form lost out—the frequency of *mutton* expressing that meaning declined. In other words, the two forms—*sheep* and *mutton*—were in competition and, in fact, were predatory on each other. When one form got used, it consumed some of the resource—the meaning—that the other also needed to be expressed; the latter lost out and was expressed less often.

There are some well-known equations from mathematical ecology that capture systems with this kind of competition—the Lotka-Volterra competition model. The model arose when, during World War I, it was noted that the number of sharks for sale in fish markets along the Adriatic coast of Italy mysteriously went up. The reason was that sea battles on the Adriatic had prevented normal fishing. Because of the decline in fishing, the number of prey fish increased, providing a greater resource for predatory fish like sharks. The fates of the prey fish and the predatory fish are linked; one provides a resource for the other.

The equations that describe this kind of system were worked out independently by Alfred Lotka in 1925 and Vito Volterra in 1926. The equations can be used to describe *interference* competition between species beyond predation, competition where two species are assumed to diminish each other's growth rate by directly interfering with each other.

In the biological system, assume that species 1 has a population size N_1 and species 2 has a population size N_2. We are interested in the rate of change in population size in the two populations and how these rates are related to each other. Each population has an inherent growth rate—r_1 for species 1 and r_2 for species 2—and the environment has an inherent carrying capacity (a maximum number that it can support) for the two species, K_1 and K_2. Without competition, the rate of change in the two populations would be characterized by the following two equations:

$$\frac{1}{N_1}\frac{dN_1}{dt} = r_1\left(1 - \frac{N_1}{K_1}\right),$$

$$\frac{1}{N_2}\frac{dN_2}{dt} = r_2\left(1 - \frac{N_2}{K_2}\right).$$

That is, the populations would grow until they reached carrying capacity, their growth rate decelerating as they did so. One might take this as a simple model of lexical spread in a population: a new word is introduced, perhaps a word for a new object, and its frequency increases rapidly until it hits a "carrying capacity" of expressing a meaning whenever speakers want to express that meaning.

However, the two species interact; the more there is of the prey species, the more resources for the predators. As the predators increase, the population of prey should drop—the predators are consuming more resources.

One needs to take into account the interaction between the two species by including *competition coefficients*: α_{12}, which describes the impact of species 2 on species 1; and α_{21}, which describes the impact of species 1 on species 2. The full system is now captured by the following equations:

$$\frac{1}{N_1}\frac{dN_1}{dt} = r_1\left(1 - \frac{(N_1 + \alpha_{12}N_2)}{K_1}\right),$$

$$\frac{1}{N_2}\frac{dN_2}{dt} - r_2\left(1 - \frac{(N_2 + \alpha_{21}N_1)}{K_2}\right).$$

That is, as the population of species 2 changes, its impact on species 1 will change (as reflected by $\alpha_{12}N_2$ in the first equation), and vice versa (as reflected by $\alpha_{21}N_1$). As the predators increase, their effect on the prey will amplify, diminishing the latter's population size. As the number of prey diminishes, this will have a negative impact on the predators, and their population should decrease. Once the predators decrease, the prey once more have a chance to renew their numbers. And so on.

The case of competition between word forms is not the same as predator-prey relations, of course. In this case, both word forms are competing for a *semantic niche*, the opportunity to express some particular meaning. Yet, some interesting analogies can be made between the Lotka-Volterra competition model and word competition to occupy a semantic niche.

The population size indicators N_1 and N_2 would correspond to the number of actual occurrences of the two word forms over some period of time. One could, for example, divide time into equal bins and count the occurrences of the two forms in each bin. Intuitively, the carrying capacities K_1 and K_2 are related to the number of occurrences of the forms that the environment can, in principle, support. This can only be estimated, but since the two word forms are competing to express the same meaning, one might estimate that

$K_1 = K_2 \approx (N_1 + N_2)$.

This is almost certainly an underestimate of carrying capacity. The intuition is that the semantic or discourse carrying capacity of a term is the frequency with which there is occasion to talk about the item denoted by the term. Since the two terms, by hypothesis, denote the same thing, one can estimate carrying capacity at least by summing their occurrences.

Now, the original system of equations simplifies to

$$\frac{dN_1}{dt} = \frac{r_1}{K_1} N_1 (K_1 - N_1 - \alpha_{12} N_2),$$

$$\frac{dN_2}{dt} = \frac{r_2}{K_2} N_2 (K_2 - N_2 - \alpha_{21} N_1).$$

Replacing K_1 and K_2 by $(N_1 + N_2)$ as an estimate of the carrying capacities, and simplifying, we get

$$\frac{dN_1}{dt} = \frac{r_1}{(N_1 + N_2)} N_1 (N_2 - \alpha_{12} N_2),$$

$$\frac{dN_2}{dt} = \frac{r_2}{(N_1 + N_2)} N_2 (N_1 - \alpha_{21} N_1).$$

The constants r_1 and r_2 represent growth constants that would tune the change in the system; these would have to be estimated by looking at real data on the competition between the words. The equations look, otherwise, quite simple.[2]

The dynamics of predator-prey competition is shown in figure 9.1. There are two sine waves, slightly out of phase with each other. The curve

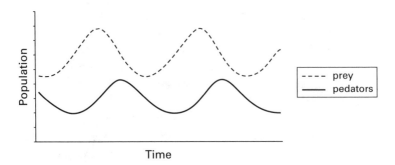

Figure 9.1
Lotka-Volterra Dynamics (From Wikipedia)

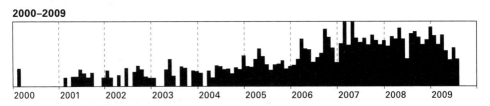

Figure 9.2
Histogram for *text-messaged* (From Google Timeline)

for the predators tracks the curve for the prey so that when the number of prey increases, the number of predators increases, for example.

In the case of two words directly competing with each other for a semantic niche, if one form increases in frequency, it must be at the expense of the other form, unless the carrying capacity—the number of times the form will have occasion to be used—also increases. Is there any evidence that this describes a real-world phenomenon?

The histogram in figure 9.2 shows how the number of occurrences of the verb *text-messaged* is distributed over time. *Text-messaging* is a verb for a new activity, sending a message in text over a cell phone. Each block in the histogram corresponds to a time bin and shows what proportion of the word's frequency occurs in that bin.

Use of the verb *text-messaged* started around the year 2000. As the activity caught on, occurrences of *text-messaged* increased with it. Notice that there are two factors here: first, the use of the word is catching on in the population of speakers, and second, the carrying capacity for the word is probably increasing because there are more occasions to talk about the activity of sending text over cell phones.

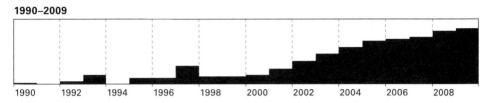

Figure 9.3
Histogram for *texted* (From Google Timeline)

Some time ago, I was walking on campus when I heard a woman say,

(15) He *texted* me last Friday.

Although I hadn't heard the word *texted* before, I knew immediately what the speaker meant: someone had sent her a text-message on her cell phone last Friday.

The words *texted* and *text-messaged* denote exactly the same activity. The two should come into direct competition for a semantic niche as soon as the carrying capacity for that meaning becomes fixed. Figure 9.3 shows the time distribution for occurrences of *texted*. Notice that *texted* seems to have had a head start over *text-messaged*, since it had occurrences throughout the 1990s. The occurrences of *texted* were, with a few exceptions, from the United Kingdom, where a service allowing text to be sent over phones was introduced at that time.

Texted really took off in the early 2000s, with the United Kingdom and New Zealand taking the lead. In the United States, people started using *texted* around 2004, and after that, as is clear from figure 9.3, its use increased steadily.

To be sure, *text-messaged* is still quite frequent, often co-occurring with *texted*. Nevertheless, comparing the timelines in figures 9.2 and 9.3, we see that *text-messaged* seems to have passed its heyday. It's interesting to compare these two figures with the curves in figure 9.1. As the resource increases—prey in figure 9.1—the predator species increases, but when the resource diminishes, the predator species does as well.

It would seem that *texted* and *text-messaged* are competing for the same resource, and *texted* is winning. As *texted* consumes more of the semantic resource—as it approaches the carrying capacity for this meaning—*text-messaged* will be edged out. This is an example of *Gause's principle*, or the *principle of competitive exclusion*, which states that if two species are too similar, they cannot coexist. Similarity is, of course, hard to measure in general; in the present case, both forms are used to express the same meaning.

The principle of competitive exclusion accords well with modeling work on the evolution of language, which suggests that while homophony—*pen* as writing instrument or as animal enclosure—is easily tolerated, synonymy—two forms with identical meanings—cannot be long supported. There are near synonymous pairs like *come* and *go*, but they are distinguished by the perspective that they take on the action; *come* involves movement toward a designated point, while *go* denotes movement away from such a point. The underlying action is the same, but the perspective on the action is sufficiently different that they occupy two different semantic niches.

But consider the fate of *sheep* and *mutton*; both originally competed for a semantic field that included both the animal and its meat. As the competition continued, the niche was split between them, with *sheep* taking one end and *mutton* taking the other. This is another example of the principle of competitive exclusion; a truce was called, and the two forms found two distinct semantic niches, both clearly derived from the original niche.

A similar account can probably be given for *cow/bull/cattle/beef* in my native dialect. The two count nouns—*cow* and *bull*—are gendered, while *cattle* is a mass noun,

(16) a. *Five cattles
  b. Five head of cattle

so there is space for a gender-neutral count noun:

(17) They slaughtered four beeves.

As expected from the principle of competitive exclusion, the various word forms distributed themselves over the semantic field and found subniches that they could occupy noncompetitively.

Will this happen with *texted* and *text-messaged*? It's hard to say. Recent technovocabulary gives us a nice testing ground. For example, consider the new use of the verbs *tweeted* and *twittered* (figure 9.4).[3] Twitter is a social web site, started in 2006, that allows people to send messages called tweets to their "followers." By 2009, the use of both *tweeted* and *twittered* was substantial. These word forms are likely in competition; the carrying capacity of this new semantic niche is still growing.

Another interesting example for study is the relative frequencies of computer terms in French. In this case, many of the words are borrowed from English, much to the horror of the Académie Française, which has tried to impose a native French vocabulary even for new activities. Thus,

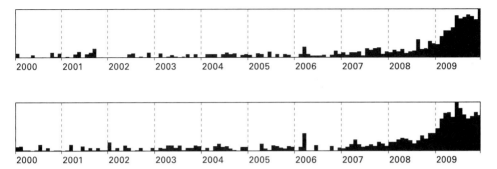

Figure 9.4
Histograms for *tweeted* (above) versus *twittered* (From Google Timeline)

a computer is *ordinateur* (or, for short, an *ordi*). A blog, however, is a *blogue*, at least for the moment. An interesting case is the translation for *email*. The official term I learned was the clumsily polysyllabic *courrier électronique*. Many French speakers never took to it, preferring *mail* or the slightly more Frenchified *mèl*. In July 2003, the Académie Française announced that the proper word, adopted from Québécois, is *courriel*.[4] With the full force of the Académie Française behind it, will *courriel* defeat *mèl*?

Finally, consider the case of *sofa* and *couch*. As far as I can tell, they mean the same thing. Both *sofa* and *couch* mean a long upholstered piece of furniture designed for several people to sit on. Now, there is a great deal of polysemy with *couch*, which has verb forms that *sofa* lacks. The two words seem to exist peacefully side by side and have done so for centuries. We would predict that they should be in competition, however. Why do they coexist so tranquilly? Perhaps they're both just too comfortable.

Semantic Hierarchies and Defaults

There is a further interesting puzzle regarding lexical games. We can see it most clearly if we consider the game of partial information for a lexical item like *pen*. The extensive game from chapter 7 is shown in figure 9.5. For both interpretations of *pen* (writing instrument or animal enclosure) the actual form *pen* has the highest utility associated with it.

Why should this be? Why does the form *pen* have so much utility relative to other forms?[5] One possible response is that *pen* is short relative to

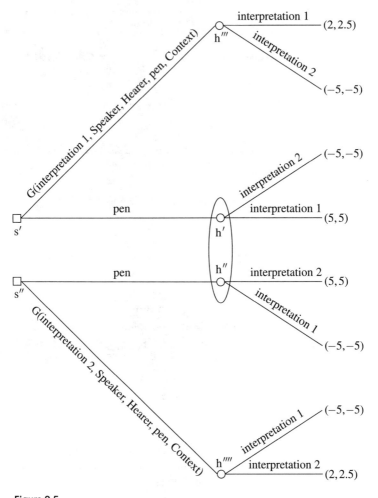

Figure 9.5
Game for *pen*

other forms. It takes less effort to say "pen" than to say "something to write with" or "an enclosure for animals."

While length may be a factor, it is unlikely to be a decisive one. There are other forms than for *pen* that might serve the same purpose:

(18) Do you have a Bic?

Bic is the brand name of a type of pen. The word is certainly no longer than *pen*, so why shouldn't we be able to use *Bic* to refer to the broader class of pens? Wouldn't this just be an instance of metonymy, using an instance of a class to refer to the broader class?

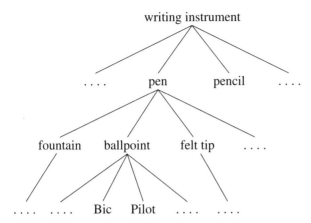

Figure 9.6
Fragment of a Semantic Hierarchy

There are cases where the name of a particular type of object comes to be the default way of referring to the entire class. For example, *Kleenex* is the brand name for a type of disposable tissue, but it has come to be the default term for referring to any type of disposable tissue. Equally, for many people, the brand name *Xerox* has become the default term for photocopy, although the Xerox Corporation has attempted to fight this use to protect its brand name. In any event, it is clear that instances of a generic category can come to be the default way of referring to that category.

Consider the subpart of a semantic hierarchy, shown in figure 9.6. The hierarchy is an *is-a* hierarchy, with a daughter being an instance of the mother, thus:

(19) a. A Bic is a ballpoint.
 b. A ballpoint is a pen.
 c. A pen is a writing instrument.

The question is, where in the hierarchy does the default term for a set lie? For example, *pen* is the default term for talking about ink-based writing instruments, while *Kleenex* is the default for disposable tissue, and *Xerox* is the default for photocopy.[6]

One approach, adopted in the prototype literature, has been to suppose that the default term for a class of objects is related to the *cue validity* and *category resemblance* of the word associated with that class. The validity of a cue is, roughly, the likelihood that knowing something is an instance of the cue reliably places it in the class:

(20) *Cue Validity*
   The validity of a cue x as a predictor of a category y is the conditional probability of y given x.

Category resemblance is defined in terms of a weighted sum of the measures of all the common features within a category, minus the distinctive features, that is, the features that belong to only some members of a given category as well as to members of contrasting categories.

Rosch (1978) wrote,

> A working assumption of the research on basic objects is that (1) in the perceived world, information-rich bundles of perceptual and functional attributes occur that form natural discontinuities, and that (2) basic cuts in categorization are made at these discontinuities. Suppose that basic objects (e.g., chair, car) are at the most inclusive level at which there are attributes common to all or most members of the category. Then both total cue validities and category resemblance are maximized at that level of abstraction at which basic objects are categorized. This is, categories one level more abstract will be superordinate categories (e.g., furniture, vehicle) whose members share only a few attributes among each other. Categories below the basic level will be bundles of common and, thus, predictable attributes and functions but contain many attributes that overlap with other categories (for example, kitchen chair shares most of its attributes with other kinds of chairs).

The idea, then, is that *pen* is a good basic-level term because it groups together all the ink-based writing implements and none of the non-ink-based implements. Notice that just saying "something to write with" fails on two counts: first, it is too general, since it includes things like pencils, and second, it is too cumbersome. Where *pen* is short, "something to write with" is wordy and requires more effort. If I need something more specific, I can fall back on a new basic-level term:

(21) Do you have a ballpoint? I need one to fill out this form.

So there's something to the notion that cue validity and category resemblance are part of the basis for apportioning utility in a language game. The idea is as follows. Suppose I want some ink-based writing implement. I maximize my chances of getting one by saying "pen" rather than "ballpoint" or "fountain pen" or "marker." Any of the latter will do for what I want, but if I say something more specific, like "ballpoint" or "Bic," I exclude members of the ink-based writing instrument category and thus reduce my chances of getting something that will serve my purposes. If my preference is to maximize my chances of getting something that will do, then I would naturally attribute more utility to *pen* than to *Bic*.

This approach is consonant with using objective frequencies as the priors for lexical access, as in chapter 7, but there is also an affinity for the ecological approach to meaning. Suppose that *pen* and *Bic* are in direct competition to occupy the default position in a semantic hierarchy. According to the reasoning just explained, *pen* has an advantage over *Bic* based on cue validity and category resemblance. It would be difficult for *Bic* to win the competition with *pen*; presumably, its growth rate is minimal, and the interference from *pen* (the α_{21} factor—or rather the $\alpha_{Bic, pen}$ factor—in the Lotka-Volterra equations) is too vast for *Bic* to replace *pen* as a default term.

But it is possible for a specific term to become the default. Thus, *Kleenex* defeated *tissue* as a default, although the latter survives and presumably is still in competition for the title. Notice that *tissue* should win on the basis of cue validity and category resemblance; nevertheless, *Kleenex* is in the dictionary.

Consider how this could work. Disposable paper tissues had been used for centuries in Japan but were a novelty in the United States when Kimberly-Clark introduced them in 1924 under the brand name *Kleenex*. Thus, the term *Kleenex* started with a clean slate—there wasn't a preexisting default term for what amounted to a new object. Thus, *Kleenex* could get a toehold in the default niche. Presumably, it was able to consume enough of the default uses to stabilize as a generic term, so that now I can offer someone a Kleenex while brandishing a box of generic tissues that I got at the drugstore.

Equally, *Xerox/photocopy/copy* have been in competition with each other for some time. The Haloid Corporation, eventually renamed the Xerox Corporation, introduced the first commercially successful photocopier, the Xerox 914, in 1959. The process used was called xerography. The Xerox 914 was an enormous success and despite its maker's, best efforts, *Xerox* became a default term that can be found in the dictionary. An interesting experiment in spatial dynamics suggests itself: Does the presence of Xerox PARC in the San Francisco Bay Area influence the spatial distribution of the word *Xerox*?

Recent technological advances provide a laboratory for the creation of default terms. Has *BlackBerry* or *iPhone* become the default term for smartphone, for example? And I often see people using MP3 players that are not made by Apple; I want to ask them whether they refer to their players as *iPods*. I'm sure that the reader is able to think of many other examples.

The question investigated in this section has some interesting philosophical implications. Goodman (1979), in his classic book on the problem of induction, proposed the predicate *grue*, which applies to all things examined before time *t* just in case they are green and to other things just in case they are blue. All observations made before time *t* that emeralds are green also confirm that emeralds are grue. Of course, after time *t*, grue things should be blue, so we know what would disconfirm the hypothesis that something is grue.

Now, you might suppose that *grue* is too complicated to work as a predicate, but many predicates involve all sorts of complex relations, for example, the relation between genes and their morphological expression. Complexity by itself won't do as a criterion. Simplicity is always a requirement, but *grue* is not outlandishly complex; in fact, *grue* is a relatively simple predicate, made slightly more complex than *green* or *blue* by adding a second predicate with a conjunction.

Part of Goodman's response to this problem was that terms like *green* and *blue* are entrenched, while *grue* isn't. But what does it mean for a term to be entrenched? Surely, Goodman intended that an entrenched expression bear sufficient social currency that it cannot easily be displaced by some other predicate. The predicate *green* is entrenched, and anything that could be called *grue* can also be called *green*; that is, *grue* is encroaching on a semantic niche already occupied by *green*. This suggests that a term is entrenched if its semantic niche cannot be invaded by another term; this is reminiscent of the definition of "evolutionarily stable strategy" from evolutionary game theory (see chapter 4).

Once a term has occupied a semantic niche long enough, any invader will face a formidable challenge in invading that niche. The interaction term—α_{ij} in the Lotka-Volterra equations—will heavily favor the home team; it will be almost impossible to topple a resident term. If a term has occupied a semantic niche long enough, the natural conservatism of speakers should work to keep the term in place. Speakers and hearers have used the term to coordinate their behavior, and any innovation would be disruptive. Innovative terms, whether *Kleenex*, *Xerox*, or *iPod*, have their best chance to occupy a niche when that part of the semantic territory is new and as yet unoccupied.

Homophones and Polysemy

There's one last issue I'd like to raise before closing. I've taken games of partial information as the basic model of strategic choice in language. The framework has a lot to offer, since it correctly captures how linguistic signals often *partially* determine an information state. There is, however, a puzzle associated with using games of partial information, one that has to do with different kinds of ambiguity in language.

Of course, in a game of partial information, all ambiguities are treated in more or less the same way. The ambiguous element maps the hearer onto an information set. *Pen* is an example of this kind of ambiguity. When I utter "pen," I could, in principle, mean either an ink-based writing instrument or an animal enclosure. Without further information from the context, or prior probabilities based on lexical frequencies, there is no way of determining what I (probably) meant.

Just about any ambiguity can be represented this way. Consider a syntactic ambiguity:

(22) Visiting relatives can be a nuisance.

The sentence in (22) can be analyzed in two different ways with two rather different interpretations: that when relatives visit, they can be a nuisance; or that the act of going to visit relatives can be a nuisance. Once again, simply uttering the sentence in (22) will leave the listener wondering which interpretation was intended, all else being equal.

One might ask whether the unified treatment accorded to ambiguities by games of partial information is appropriate in all cases. I think a case can be made that, in general, the unified treatment is to be preferred over a more disparate account. There is, however, a type of ambiguity that seems to require special treatment, if only because games of partial information may be rather a blunt instrument for dissecting such instances.

Let's return to the ambiguity of *pen* that started the whole discussion. The various meanings of *pen* (writing instrument or animal enclosure) are associated with a single phonological form via a chain of historical happenstance. The writing instrument version of *pen* is ultimately derived from Latin *penna* 'feather', a relic, no doubt, of the use of quills. The sense of *pen* associated with an animal enclosure is from Old English *penn*, a word of unknown origin. It is surely just an accident that the word for writing instrument wound up sounding identical to the word for animal enclosure.

This type of ambiguity is *homophony*, where two unrelated forms come to sound alike because of regular sound changes or historical serendipity. But there is another type of ambiguity, *polysemy*, that seems to have different properties. I've already touched on the distinction when discussing the different meanings of *crawl*. In the case of *crawl*, the different meanings seem less serendipitous.

Consider, for example, the word *mouth* in English. It can mean the opening in the lower part of the face, but it can also mean any opening to a hollow structure, for example, the mouth of a cave or the mouth of a bottle. It can also mean the entrance of a harbor or the place where a river meets the sea. These different senses do not seem accidental in the way that the two different meanings of *pen* do.

It's useful here to compare languages. Suppose I am discussing farming techniques with a monolingual French farmer. I can't think of the proper French word for *pen* in the sense of animal enclosure, so I simply take a translation of *pen* (writing instrument) and say something like "stylo" or "plume," either of which can translate to "writing instrument," and I bumble out with,

(23) Est-ce que vous avez un stylo pour vos porcs?
     is-it that you have a pen for your pigs
     'Do you have a pen (writing instrument) for your pigs?'

No doubt I would be met with incredulity and puzzlement. Why on earth would I want to know whether his pigs have something to write with? Of course, it would never occur to me to try to pull off such a bad translation, since I would never suppose that the homophony in English would carry over to French.

Things are rather different when it comes to polysemy. I might, for example, try to point out the mouth of a cave by saying "la bouche (mouth) de la grotte." This is incorrect—I should use *entrée* for the opening of a cave—but I'm unlikely to be met with incomprehension. I've still made an error but one that seems qualitatively different from my error with *pen*.

The homophony of *pen* is pure serendipity, while the polysemy of *mouth* is less so, but this is also not some necessary fact about language. Things could have ended up differently. French diverges from English in the way it divides things up: humans and volcanoes have a *bouche*, caves have an *entrée*, rivers have an *embouchure*.

I would submit that the difference between homophony and polysemy is one that linguists should take a profound interest in. Kripke (1977) makes the following point:

> We thus have two methodological considerations that can be used to test any alleged ambiguity. "Bank" is ambiguous; we would expect the ambiguity to be disambiguated by separate and unrelated words and some other languages. Why should the two separate senses be reproduced in languages and related to English? First, then, we consult our linguistic intuitions, independently of any empirical investigation. Would we be surprised to find languages that used two separate words for the two alleged senses of a given word? If so, then, to that extent our linguistic intuitions are really intuitions of a unitary concept, rather than of a word that expresses two distinct and unrelated senses. Second, we can ask empirically whether languages are in fact found that contain distinct words expressing

the allegedly distinct senses. If no such language is found, once again this is evidence that the unitary account of the word or phrase in question should be sought. (Kripke 1977, 19)

I would argue for a more nuanced approach. I'm not surprised that French treats the opening of a cave differently from the opening in the lower part of the face, but I wouldn't be surprised if it did behave like English (it agrees with English about volcanoes and human faces).

There is some evidence from neuroscience that the brain treats polysemy differently from homophony as well. Pylkkänen, Llinás, and Murphy (2006) looked at just this question. We know that homophones involve two distinct lexical entries that happen to have identical phonological realizations attached to them; thus, they should behave like two independent words. The researchers assumed that the different senses of a polysemous item would be listed in the same entry—the *shared root hypothesis*. Seeing a polysemous word in two different senses should activate the same lexical entry, revealing repetition effects. This repetition effect should be absent in the case of homophonous pairs, since they are, by hypothesis, different lexical entries.

The experiment involved a priming paradigm; priming happens when one element facilitates a response to another element. For example, suppose I am asked to decide whether a letter sequence that I'm looking at is a word. In one condition, I might first be exposed to a word like *cup*. It would take me some amount of time to respond that it is a word (usually a button push is used for the response). Then I might be shown a word like *shoe*. Since the two words are unrelated, the responses are not expected to differ in the two cases. Now suppose that I am first exposed to a word like *doctor*. Then I am exposed to *nurse*. Here, a semantic neighborhood has been activated. If my response time was faster for *nurse* than for *shoe*, we would say that *doctor* primed *nurse*.

In the experiment by Pylkkänen et al., subjects were asked whether a targeted two-word phrase made sense. They were shown a series of two-word phrases. The first phrase was the prime that was intended to facilitate the response; the second phrase was an unrelated prime (actually, a neutral condition); and the final phrase was the target, the phrase they wanted to test. Examples are shown in table 9.1.

In the homophony case, responses to homophonous pairs like *bank* were compared. The two-word phrase disambiguated the sense intended. In the polysemy case, two senses of a polysemous word like *paper* (the material or a publication) were tested. Finally, semantically related pairs were tested.

Table 9.1
Stimuli for an Experiment on Priming

	Related Prime	Unrelated Prime	Target
Homophony	river bank	salty dish	savings bank
Polysemy	lined paper	military post	liberal paper
Semantic	lined paper	clock tick	monthly magazine

Source: Pylkkänen, Llinás, and Murphy (2006).

The researchers first tested subjects in a behavioral experiment, that is, they presented the subjects with triples and tested the speed with which subjects made their judgments. Here, they found that using a word in the same sense twice led to a faster response than switching senses. This occurred for both homophones and polysemes, so no interesting difference was noted.

Things changed, however, when electrophysiological responses were measured. The researchers used magnetoencephalography (MEG) to measure magnetic fields produced by electrical activity in the brain. MEG measurements have fairly good spatial resolution and excellent temporal resolution; in other words, one can get a fair idea of where in the brain things are happening and an excellent idea of when things are happening.

Pylkkänen and colleagues reasoned that if polysemy involves distinct sense and root sharing—that is, different sense but one lexical entry—they should get a response about 350 milliseconds (so-called M350) after presentation that would show a particular pattern distinct from homophones, where the sense is unrelated, and from semantically related primes. If polysemes were represented like homophones—that is, different sense and different lexical entries—the response at M350 should be explained as a combination of the effects found in homophones plus semantic priming.

In other words, they were looking for a unique response at M350 for polysemous pairs, a response different from the response found in homophony and from the response found in semantic priming. Further, this unique response should not be explicable in terms of the combined effect of phonological similarity (homophony) and semantic similarity (semantic priming).

The exact details of the experimental results are of less interest here than the fact that the researchers showed that there is a difference in neurophysiological response to polysemes as opposed to homophones.

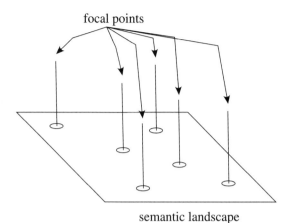

Figure 9.7
Semantic Landscape with Focal Points

We needs to account for the fact that the brain can sort items into two different classes, one for homophones and the other for polysemes.

This section has actually outlined the beginnings of such an account. It developed the idea that terms are associated with focal points in a semantic landscape. So, for example, *dog* would hit the focal center of a domestic canine space—friendly, furry creatures that bark, and so on. The meaning in context can be tuned via a metric—what is taken in the context to be a dog—or by modifiers that can shift the focal center of the term, as in *pet fish*.

Homophones would be listed multiple times in the lexicon, each with its own focal center. These focal centers could be widely separated in the semantic landscape; a writing instrument and an animal enclosure have little to do with each other, so one wouldn't expect the focal centers for the different readings of *pen* to be adjacent in semantic space. Each focal sense would be associated with a distinct lexical item; the items themselves just happen to sound alike.

Let's consider how to treat polysemes. Figure 9.7 shows a schematic drawing of a semantic landscape with a set of focal points distributed over it. Assume that the distribution of focal points is random, the result of various historical processes, cultural practices, and so on. Of course, not every point in the space will be a focal point; in fact, most points will not.

Because the space is sparsely covered by the focal points associated with lexical items, there is an interesting packing problem. The goal is to be able to signal any point in the space to another speaker using some term in the lexicon. Any focal element (the combination of a term plus its focal point) in the space can be thought of as entering into competition for semantic niches with any adjacent focal element. That is, adjacent focal elements should compete for territory in the semantic landscape, where *territory* indicates the right to occupy a semantic niche at that point.

The speaker and the hearer are involved in a coordination game where a term (and its associated focal point) is used to signal a point in the semantic space. The speaker is faced with the strategic problem of which term to use to signal the intended point; the hearer is faced with the strategic problem of picking the intended point in semantic space given that the speaker has used some particular term. Since the space is sparsely covered by focal elements, the problem is not trivial. However, through repeated plays of these coordination games in a community, conventions begin to emerge. These conventions might vary from community to community.

To take a particular example, a word like *mouth* might begin as denoting the opening in the lower part of an animal's face and extend itself to other openings, points that are nearby in the semantic landscape. Thus, animals, caves, volcanoes, bottles, rivers, guns, and bays all acquired mouths as *mouth* spread from openings to entrances. In French, *bouche* 'mouth' came into competition with *entrée* 'entrance', and the battles came out with a different division of the territory; animals and volcanoes have a *bouche* but caves have an *entrée*. It turns out that rivers have an *embouchure*, and so on. The resulting division of territory is a contingent fact of history, but not entirely serendipity.

Thus, *sheep* and *mutton* began at the same focal point. Competition between them led to a division of the territory: *sheep* denotes the animal, *mutton* denotes the meat (and sometimes the animal destined to be meat; I suppose that sheep raised just for their wool are not muttons).

Frequent words, particularly words that have been in the language a long time, might be expected to be polysemous. According to Britton (1978), 44 percent of words drawn randomly from the dictionary are polysemous, and of the hundred words most frequently used in English, ninety-three have more than one meaning. Lee (1990) showed that more frequent words were more likely to be polysemous and that the length of time that the word has been in the language accounts for much of the variance in polysemy. This accords well with the lexicographer's intuition that the frequency of a word and the length of time it has been in the language are important factors in its polysemy. All of this works well with the Lotka-Volterra model of competition. As noted, entrenchment should reinforce the use of a word; the longer the word is in the language and the more frequently it is used, the more chances it has to compete with other terms and gain territory. Of course, there are competing forces. In addition to the tendency to acquire more adjacent meanings in the semantic space, there is the need for a certain amount of clarity. If I used the monosyllable "uh" to signal all my intended meanings, the hearer would be faced with an insurmountable problem of coordinating with me. Communication would fail; the monosyllable approach would be a poor strategy to follow.

This treatment of polysemy, then, involves the same item's being associated with a larger territory in semantic space. This could be expressed as the same lexical item's being associated with more than one focal point in the space. This is exactly the shared root model that Pylkkänen, Llinás, and Murphy tested in their experiment.

Notice that the analysis of polysemy here is rooted in coordination games—the territories are being carved out by speakers and hearers coordinating their behavior. In both polysemy and homophony, choices must be made. In the case of homophony, the grammar offers two lexical items that sound alike. In the case of polysemy, the grammar offers a single lexical item with more than one focal sense. Speakers and hearers must make choices in either event, but the etiologies of the choices have different analyses. Homophony is ultimately the result of accident; polysemy is the contingent result of economic and ecological forces. Thus, games of partial information can represent the choices that speakers and hearers face in these cases. Bear in mind, though, that the choices are the result of a variety of historical forces.

Into the Artificial World

We now have a general framework for thinking about lexical meaning and the evolution of conventions. Some aspects of the model should be straightforward to test empirically. For example, we should be able to see the competition between items like *texted* and *text-messaged*. All that is required is access to a sample of texts from the Internet—blogs would be ideal—over a period of time. One could take the data, bin them into time slots (say, slots of one month each), and then count the occurrences of the competing items.

I've provided an illustration of this method with histograms that show what proportion of a term's use occurs in each bin (see figures 9.2–9.4).

While this is informative, the real test would come from the actual numbers. Nothing stands in the way of testing this aspect of the model except access to the raw numbers.

There are other aspects of the model, though, that are not feasible to test using conventional corpus linguistics. The useful fact about the competition between *texted* and *text-messaged* is that instances of the different forms can be counted. While the resulting counts may not perfectly reflect the underlying semantics of the terms, given a large enough sample the counts should do.

Things are less clear with regard to the competition for default position in a semantic hierarchy or the conquest of semantic territory found in polysemy.

Consider my favorite example, *pen*. Even assuming that we had sufficient data from the relevant period—when *pen* was competing with other forms to become the default—we would have to decide for each instance whether the use was intended to refer to a specific pen or to pick out a class of writing instruments. Equally, if we were trying to study the evolution of polysemy, we would have to decide, for each use of the term, what sense the term was being used in. Even if we had sufficient information from the texts, it would be incredibly labor-intensive to mark each occurrence for sense.

There may, in fact, be a principled limit on the use of historical corpora—collections of texts—in the study of sense change. Advances in computational linguistics suggest there soon will be robust algorithms for automatically marking words for their sense. Suppose we take a sample of English from the eighteenth, nineteenth, and twentieth centuries, break it into bins by decades, and try to distinguish the sense of content words. Zipf's law[7] guarantees that the frequency of occurrence of almost all content words will be so low that we will not be able to carry out a useful statistical analysis on them. Because of the sparse distribution of these words, we will not be able to see their behavior in the corpus with any reliability.

Thus, it is necessary to seek a different method for exploring the question at hand, namely, the evolution of conventions regarding linguistic defaults and the development of polysemy. I have proposed that both phenomena spring from the same underlying mechanism and that the responsible mechanism involves social convention rather than purely psychological mechanisms. Surely, the problem is of sufficient interest to warrant a rigorous investigation.

The obvious course in this circumstance is to use simulations to test understanding of the essential properties of the problem. One could use evolutionary game theory to develop a population-based simulation of default selection and polysemy. Recall, however, that evolutionary game theory supposes that the population consists of simple, that is, irrational, agents. These agents have the chance to play a single strategy, after which fitness is calculated and the next generation of the population is constructed on the basis of the fitness calculation.

This approach has a curious result. Recall the discussion in chapter 4 about the hawk-dove game, where animals interact for territory; the territory is of value v, and if the animals fight, there is a cost c to the animals. We saw that if $v > c$, then the game reduced to a prisoner's dilemma game. The result was that playing hawk—the aggressive strategy—dominated the dove option. Because of this, a population of hawks cannot be invaded by doves; hawk is the evolutionarily stable strategy (ESS).

Contrast this with the result from the iterated prisoner's dilemma game. When players play Prisoner's Dilemma with each other repeatedly, they tend to cooperate, that is, they behave like doves. At first, this appears like a paradoxical outcome. Why should evolutionary game theory predict that hawk (defect) is the ESS, while the theory of repeated games tells us that dove (cooperate) is the equilibrium? The obvious explanation is that the individual agents in the evolutionary game-theoretic setting are not playing a repeated game—they get to play only once—so they have no memory; while the population learns, the individuals do not.

With repeated games, the situation is quite different. In this case, reputation matters—if one has the reputation of a defector, then the opponent is more likely to defect—and learning is not only possible but necessary. I suspect that learning and strategic choice make all the difference. While I argued in the previous section that the process of default selection and polysemy was necessarily social, requiring the members of a population to engage in strategic coordination, the individual agents are boundedly rational. They have a memory and make choices to the best of their ability. The model that takes agents as irrational will necessarily distort the outcome of the process.

If this line of reasoning is correct, then *agent-based modeling* should be used to explore the social evolution of linguistic defaults and polysemy. In this type of modeling, the agents can alter their choices on the basis of experience; they can modify their behavior on the basis of past experience.

The essential idea behind a simulation of default selection is the following. Start with a population of agents that will be paired off to play a coordination game. One agent will transmit a signal, and the other agent will pick an object from a model on the basis of the signal. If the first agent is satisfied with the second agent's choice, the payoff to both agents

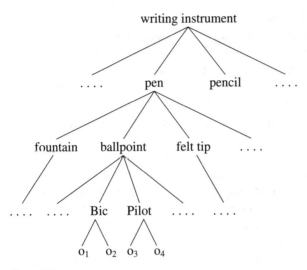

Figure 9.8
Model Organized by a Semantic Hierarchy

will be positive; if the first agent is not satisfied, then miscommunication has occurred, and there is either no payoff or a negative payoff.

I represent the situation with the simple model of writing instruments discussed earlier. Suppose there is a model consisting of a finite set of objects, all of them writing instruments in this case. These writing instruments, o_1, o_2, and so on, can be organized according to the semantic hierarchy for writing instruments (figure 9.8).

Suppose there is a set of tasks that can be accomplished with writing instruments; the twist is that each task can be accomplished by different kinds of writing instruments. Writing a check needs a pen—fountain, ballpoint, roller, or felt tip—while making a carbon copy requires a ballpoint. Doing arithmetic calculations would require a pencil; drawing might need charcoal or a colored pencil. One could define a whole series of tasks that can be best accomplished with one kind of writing instrument or another.

Now reach into the population of agents and pair them off for a game. One agent is designated the speaker, and the other is designated the hearer. The speaker is given a task that only some objects in the model can satisfy. The hearer is given access to some but not all of the objects in the model. Both agents can see the model and have access to the semantic hierarchy, but only the speaker knows the task he has been assigned, and only the hearer knows which objects she has access to.

The speaker now transmits a signal to the hearer. The signal is one of the node labels in the semantic hierarchy. The hearer takes the signal and tries to give the speaker what he wants, supposing that she has access to such an item. If the speaker gets an object that satisfies the requirements of his task, then both the speaker and the hearer get a positive payoff. If the speaker fails to get an object that satisfies his task requirements, then the speaker, at least, gets a negative payoff, and the hearer gets nothing.

Suppose that the speaker is given the task of writing a check, for which he needs ink. He uses the root node of the hierarchy, *writing instrument*, and receives for his pains a charcoal pencil, completely unsuitable for writing a check. Over a population, this kind of scenario will be frequent enough to drive speakers down the hierarchy.

Suppose, on the other hand, the speaker calls for a Bic. If the hearer has one, she will give it to him, but her odds of having one are, in general, lower than her odds of having something that will count as a pen, which is what the speaker needs. Over a population, we should see speakers moving higher in the hierarchy in this kind of case.

In fact, the node in the hierarchy that maximizes the speaker's chance of getting something good for writing a check is *pen*. This is exactly what we would expect from cue validity (see example (20)). Once *pen* has won the contest as a linguistic default, it should be very difficult for a pretender like *Bic* to invade.

However, metonymy sometimes wins the day. All else being equal, this will happen when a particular element dominates the type in the semantic hierarchy; for example, Kleenex was the only paper tissue back in the 1920s. Asking for a Kleenex would have worked as well as asking for a tissue. Thus, *Kleenex* had a chance to become entrenched as the linguistic default for tissues.

Since the agents in this model survive multiple rounds of play, they have a chance to learn from their past behavior. This is a key difference between agent-based models and evolutionary game theory. We could try out a variety of learning rules and allow agents varying degrees of memory.

We could also place the agents on a lattice and let them only play within their neighborhood. This would allow the modeling of spatial properties as well as social networks. For instance, perhaps all the high-status agents use iPods. I might refer to my off-brand MP3 player as an iPod just for some reflected glory.

This suggests that we could manipulate the payoffs of the game. We might give a slightly higher payoff for using shorter terms; or we could add a bit of utility to the payoff for using high-end words like *iPod*. We could explore a variety of payoff functions to see how they altered the outcome of the game for the population.

How would we evaluate the results? Presumably, each agent would learn a strategy profile. The profile would include moves like

(write a check, *pen*),

where the first element is a task and the second element is a signal. Another move might be

(*pen*, o_4),

where the first element is a signal and the second element is an object that the agent has access to.

We could then let the agents play the game for some sufficiently large number of iterations. At the end, we could gather statistics on the strategy profiles that the agents have developed. We should then be able to see whether any conventions have emerged in the population.

A similar type of simulation could be used to study the development of polysemy in a population. In this case, we would need to build a semantic landscape where objects were arrayed in a space according to their feature properties. The exact features would depend on the space we were trying to model.

Objects would be arrayed over the landscape according to their feature similarity. As a result, objects would cluster together into clumps of similar objects. Presumably, the centers of such clumps would be likely candidates for focal points in the semantic space. Each focal point would then be assigned a distinct term.

Once again, the players would engage in a game of coordinated signaling. One agent would play the speaker, while the other would play the hearer. Some subset of the objects in the space would be selected for a local model. The speaker would be assigned one of the objects in the local model. His task would be to strategically select a term that would get the hearer to pick the correct object from the local model. If the speaker and the hearer can coordinate around the same object, then they get a positive payoff.

This method could be used to study the beginnings of coordinated reference as well as the problem of how vocabulary items pack themselves over a semantic space. I would expect that the various solutions to these packing problems would show the beginnings of polysemy.

The approach examined in this chapter builds on the idea that linguistic meaning is largely social. My hypothesis is that meaning arises out of social coordination games. While I have no doubt that speakers and hearers have mental representations that they use in the computation of meaning, it is the content of these representations that is ultimately crucial to them and that makes a difference in understanding meaning. Ultimately, this content is a public matter, not subject to private delegation but worked out through social practice. Linguists must engage the social and strategic side of meaning.

Further Reading

This chapter brings us back to the puzzle of reference introduced in chapter 1. Searle's (1969) speech act theory of reference is helpful. The problem of reference is not simply a philosophical puzzle; it is central to understanding language and therefore of concern to working linguists. The solution to the problem surely rests on working out what is meant by *concepts*, since concepts mediate between language and the world. The literature on concepts is vast and quite complicated. Fortunately, Murphy (2002) provides a useful and well-organized guide to the problems. I could not have hoped to navigate the dangerous waters broached in this chapter without it.

The Aristotelian approach to concepts is basically that of Tarski (his early paper on truth is published in Tarski 1983) and his student Richard Montague (1974). The approach seems to represent conventional thinking in much of natural language semantics.

Prototype theory is most associated with Eleanor Rosch; see, for instance, Rosch (1978) or Rosch and Mervis (1975). Armstrong, Gleitman, and Gleitman (1983) give some cogent but nonfatal criticisms of prototype theory. In linguistics, prototype theory is closely associated with Lakoff and Johnson (1980). Taylor (1995) gives a very useful overview of prototype theory and linguistic categories.

Schelling's (1960) focal points and social coordination are a good way to think about prototypes and very close to what Rosch had in mind, as the title of her 1975 paper "Cognitive Reference Points" attests. See also Lewis (1969). Mehta, Starmer, and Sugden (1994) performed some careful experiments to show that coordination is the crucial factor. Sugden (1995) provides an analysis of focal points. Sugden and Zamarrón (2006) give a careful discussion of Schelling's idea. Bardsley et al. (2010) describe empirical work on focal points. Sally (2003) notes the importance of this empirical work for understanding linguistic meanings and implicature.

Labov's work on vagueness and typicality is reported in Labov (1973); see also Taylor (1995). Vagueness, in general, is thoroughly discussed in Williamson (1994); my own preference is that vague terms have no conventions regarding their limits. The Sorites paradox is only a paradox in that it presupposes a convention about where heaps begin and end.

Van Deemter's (2010) book is an invaluable addition to the literature on vagueness and extremely readable. Pietroski (2005) treats Austin's (1975) example, "France is hexagonal," as a variety of vagueness. Pietroski's discussion is an interesting attempt to put aside the standard truth-conditional account, grounded in the Tarskian approach, while maintaining many features of that approach; in particular, while I hold, following the tradition of Wittgenstein and Austin, that meaning is usage, Pietroski seems to see meaning as originating in deep cognitive systems that are necessarily prior to usage.

The Lotka-Volterra model of competition is discussed in Sigmund (1993), Hofbauer and Sigmund (1998), Kot (2001), and Nowak (2006). Clark, Parikh, and Ryant (2007) propose that the Lotka-Volterra model could be used to understand default selection in a semantic hierarchy. On the exclusion principle, see May (1973). In general, the idea that words spread like a contagion is suggested by Mackay (1969); I don't think he uses the word *disease*, however.

Cue validity is discussed in Tversky (1977) and *grue* in Goodman (1979).

The puzzle of homophones and polysemy has been neglected; it's hard to find good discussions of it. Taylor's (1995) book is an exception. Lyons (1977) also discusses the issue. Pylkkänen, Llinás, and Murphy (2006) show that polysemy and homophony are recognized as different by the brain.

Agent-based modeling may be the only way to understand the processes thoroughly, particularly if my intuition about the consequences of Zipf's law is correct. The Miller and Page (2007) discussion of agent-based models is helpful, as is Epstein's (2006). If I'm right here, we can think of meaning as an emergent property of social systems.

As usual, Schelling is ahead of the game. Schelling's (1978) book is a primer on emergent properties of social systems. He described the first agent-based simulation in his paper "Dynamic Models of Segregation" (collected in Schelling 2006); using a checkerboard and some simple rules, he simulated the dynamics of segregation.

Notes

Preface

1. Which, of course, you're not holding now as I write this. Furthermore, I'm not writing this now as you're holding it. Nevertheless, you (in the future) know exactly what I (in the past) mean.

2. My own inclination is to treat grammar not as an object but as a cloud of social practices that we use to coordinate our behavior. My intent in this book is to lay the foundations for this approach.

Chapter 1

1. We paid $5.00 a day, which was considered a generous wage for the 1960s. As of 2005, the average daily wage for a worker in a *maquiladora* (one of the assembly plants that have sprung up along the border) was $5.00 to $6.00 a day, according to the New York State Labor-Religion Coalition. Currently, a domestic worker can get $18.00 to $23.00 a day, according to an article in the *Frontera NorteSur* online. The minimum wage in Juárez is still 52 pesos a day, about $5.20, although with bonuses one can get up to 125 pesos, according to MESA, the Movement for a Solidarity Economy in the Americas.

2. The following example owes a great deal to Lila Gleitman. I am, of course, coopting the example for my own purposes, so you shouldn't blame her for any of my conclusions.

3. We'll return to the problem of Claude in chapter 9; suffice it to say that I've played a trick by making you aware of Claude's peculiar biography. The real question is how we would come to categorize tigers and other natural (and artificial) things.

Chapter 2

1. Real proof theory would look much more mathematical than what I've shown in (10), but since this isn't a book about proof theory, I just want to convey an idea of how it works.

2. Much less, human intelligence. Perhaps one could grant that the computer simulation was genuine intelligence, but an intelligence that differed somehow from human intelligence.

Chapter 3

1. There is only one difference: On Earth, water is made up of H_2O, while on Twin Earth it is made up of a chemically distinct compound XYZ, which amazingly looks, tastes, and works just like water.
2. The force of a word might diminish from overuse. It's not clear that this process is analogous to monetary inflation.

Chapter 4

1. I should note that *cooperation* receives a very special interpretation in game theory. A game is cooperative when one of the player's choices is constrained by a binding agreement, like a contract. I do not use *cooperation* in that sense here; rather, I use it to mean something likes its ordinary sense of harmonized, collaborative behavior.
2. In honor of Satyajit Ray's 1968 film *Goopy Gyne Bagha Byne*, distributed as *The Adventures of Goopy and Bagha*.
3. See a transcript at ⟨http://people.ischool.berkeley.edu/~nunberg/authenticity.html⟩.
4. Game theory mavens will recognize this game as the battle of the sexes game or the Bach or Stravinsky? game familiar from introductory game theory texts.
5. Albert W. Tucker is credited with giving the game its prisoner's dilemma interpretation and name.
6. Unless, of course, one of them is driving an SUV!
7. There is a problem with Tit for Tat strategies: Tit for Tat will cycle between cooperate and defect if it plays another Tit for Tat–like strategy that played defect once. This last strategy might be one that experiments with an occasional defection. A more robust strategy might try to find a way out.
8. Notice that both players presumably prefer the Pareto-dominant payoff; they can get the preferred outcome if each player has the certainty or assurance that the other player will choose the best outcome. For this reason, this class of games is often called *assurance games*.
9. Or worse. "The problem here is that there will always be some uncertainty about how quickly he can acquire nuclear weapons. But we don't want the smoking gun to be a mushroom cloud," said the National Security Advisor Condoleeza Rice, to justify the invasion of Iraq (interview, September 8, 2002; see ⟨http://transcripts.cnn.com/TRANSCRIPTS/0209/08/le.00.html⟩). It would seem that our political leaders still hold the dark view informed by Prisoner's Dilemma.

Notes

Chapter 6

1. Their use is sensitive to a variety of properties of the discourse context; in fact, a full account of their meaning is a good example of the use of game theory in the analysis of linguistic meaning.

2. In fact, he was. In his youth he had lovely curly brown locks, but he became bald early. He employed over forty wigmakers and always wore a wig. Unsurprisingly, wigs became the fashion in Louis's court.

Chapter 7

1. Actually, my dictionary lists more than five different meanings for *pen*, including the two I mention in the text. *Pen* can also denote a female swan, a penitentiary, the internal shell of a squid, and an acronym for the International Association of Poets, Playwrights, Editors, Essayists, and Novelists. And let's not forget the homophonous *Penn*, naming my fair institution.

2. The speaker again violates the manner maxims and neatly implicates that he, the speaker, is a pretentious blowhard.

Chapter 8

1. We might argue that Amy is also violating the manner maxim. Her method of getting me to go to the meeting is not perspicuous. This is related to the question of politeness, addressed later.

2. I've changed the names of the terms in the equation given by Brown and Levinson (1987) to make the terms more transparent.

Chapter 9

1. Many early naturalists thought that the platypus was a hoax. Some believed that the body had been produced by an Asian taxidermist.

2. In fact, the right system is probably an *exploitation* competition model, since the forms are competing to exploit a common resource. I have not yet worked out this system of equations, so I will stick with the more basic model.

3. Thanks to Jon Stevens for this example.

4. Thanks to Gillian Sankoff for pointing out *courriel* to me.

5. I acknowledge Bill Labov for insisting on the importance of this question.

6. At least in some places. When I presented some of this material at Stanford University, many people objected that *Xerox* was not the default for photocopy and that they had never heard anyone use *Xerox* as a default for photocopy. I take them at their word, of course, but I note the proximity of Xerox PARC, plus the stated desire of the Xerox Corporation to control its brand, as an explanation for the discrepancy.

7. Zipf's law states that, given a corpus, the frequency of any word is inversely proportional to its rank. Thus, the most frequent word will occur about twice as often as the next most frequent word.

References

Abelard, Peter. 1922. *Historia calamitatum: The story of my misfortunes*. Translated by Henry Adams Bellows. Saint Paul, MN: T. A. Boyd. Originally written ca. 1135.

Armstrong, Sharon Lee, Lila R. Gleitman, and Henry Gleitman. 1983. What some concepts might not be. *Cognition* 13:263–308.

Austin, John L. 1975. *How to do things with words*. 2d ed. Cambridge, MA: Harvard University Press.

Axelrod, Robert. 1984. *The evolution of cooperation*. New York: Basic Books.

Axelrod, Robert, and William D. Hamilton. 1981. The evolution of cooperation. *Science* 211(4489):1390–1396.

Bardsley, Nicholas, Judith Mehta, Chris Starmer, and Robert Sugden. 2010. Explaining focal points: Cognitive hierarchy theory versus team reasoning. *Economic Journal* 120(543):40–79.

Barwise, Jon. 1989. *The situation in logic*. Stanford, CA: CSLI Publications.

Barwise, Jon, and John Etchemendy. 1987. *The liar: An essay on truth and circularity*. Oxford: Oxford University Press.

———. 1989. Model-theoretic semantics. In *The foundations of cognitive science*, ed. Michael I. Posner, 207–243. Cambridge, MA: MIT Press.

———. 2002. *Language, proof and logic*. Stanford, CA: CSLI Publications.

Barwise, Jon, and Lawrence Moss. 1996. *Vicious circles: On the mathematics of non-wellfounded phenomena*. CSLI Lecture Notes No. 60. Stanford, CA: CSLI Publications.

Barwise, Jon, and John Perry. 1983. *Situations and attitudes*. Cambridge, MA: MIT Press.

Beaver, David I. 1997. Presupposition. In *Handbook of logic and language*, ed. Johan van Benthem and Alice ter Meulen, 939–1008. Cambridge, MA: MIT Press.

———. 2001. *Presupposition and assertion in dynamic semantics*. Stanford, CA: CSLI Publications.

———. 2004. The optimization of discourse anaphora. *Linguistics and Philosophy* 27(1):3–56.

Beaver, David I., and Henk Zeevat. 2007. Accommodation. In *The Oxford handbook of linguistic interfaces*, ed. Gilliam Ramchand and Charles Reiss, 503–538. Oxford: Oxford University Press.

Benz, Anton, Gerhard Jäger, and Robert van Rooy. 2006. An introduction to game theory for linguists. In *Game theory and pragmatics*, ed. Anton Benz, Gerhard Jäger, and Robert van Rooy, 1–82. Palgrave Studies in Pragmatics, Language and Cognition. Basingstoke, UK: Palgrave Macmillan.

Bever, Thomas G. 1970. The cognitive basis for linguistics structures. In *Cognition and the development of language*, ed. John R. Hayes, 279–362. New York: Wiley.

Binmore, Ken. 2007. *Game theory: A very short introduction*. Oxford: Oxford University Press.

Blackburn, Simon. 2005. *Truth: A guide*. Oxford: Oxford University Press.

Borel, Émile. 1921. La théorie du jeu et les equations intégrales à noyau smétrique. *Comptes Rendus Hebdomadaires des Séances de l'Académie des Sciences* 173:1304–1308. Translated by Leonard J. Savage as "The theory of play and integral equations with skew symmetric kernels," *Econometrica* 21(1953):101–115.

Bostock, David. 1997. *Intermediate logic*. Oxford: Oxford University Press.

Bourdieu, Pierre. 1991. *Language and symbolic power*. Cambridge, MA: Harvard University Press.

Brandom, Robert B. 1994. *Making it explicit: Reasoning, representing and discursive commitment*. Cambridge, MA.: Harvard University Press.

Breheny, Richard. 2002. Pragmatic analyses of anaphoric pronouns: Do things look better in 2-D? Manuscript, RCEAL, University of Cambridge.

Britton, Bruce K. 1978. Lexical ambiguity of words used in English text. *Behavior Research Methods and Instrumentation* 10:1–7.

Brown, Penelope, and Stephen C. Levinson. 1987. *Politeness: Some universals in language usage*. Cambridge: Cambridge University Press. Based on original material in *Questions and politeness*, ed. Ester N. Goody (1978), with corrections, new introduction and bibliography.

Bryan, Michael F. 2004. Island money. Federal Reserve Bank of Cleveland, Economic Commentary.

Camerer, Colin. 2003. *Behavioral game theory: Experiments in strategic interaction*. Princeton, NJ: Princeton University Press.

Carlson, Lauri. 1983. *Dialogue games: An approach to discourse analysis*. Dordrecht, Netherlands: D. Reidel.

Carnap, Rudolf. 2003. *The logical structure of the world and pseudoproblems in philosophy*. Chicago: Open Court.

Carroll, Lewis. 1946. *Alice in wonderland and through the looking glass*. Illustrated Junior Library. Kingsport, TN: Grossett and Dunlap. Originally published 1865 and 1871.

Chomsky, Noam. 1986. *Lectures on government and binding: The Pisa lectures*. Dordrecht, Netherlands: Foris Publications.

Clark, Herbert H. 1992. *Arenas of language use*. Chicago: University of Chicago Press and Center for the Study of Language and Information.

———. 1996. *Using language*. Cambridge: Cambridge University Press.

Clark, Robin. 2007. Games, quantifiers and pronouns. In *Game theory and linguistic meaning*, ed. Ahti-Veikko Pietarinen, 207–227. Vol. 18 of Current Research in the Semantics/Pragmatics Interface. Amsterdam: Elsevier.

———. 2009. Games, quantification and discourse structure. In *Games: Unifying logic, language, and philosophy*, ed. Ondrej Majer, Ahti-Veikko Pietarinen, and Tero Tulenheimo, 139–150. New York: Springer.

Clark, Robin, and Murray Grossman. 2007. Number sense and quantifier interpretation. *Topoi* 26:51–62.

Clark, Robin, and Prashant Parikh. 2007. Game theory and discourse anaphora. *Journal of Language, Logic and Information* 16:265–282.

Clark, Robin, Prashant Parikh, and Neville Ryant. 2007. Evolving linguistic defaults. Manuscript, University of Pennsylvania.

Cowie, Fiona. 1999. *What's within? Nativism reconsidered*. Oxford: Oxford University Press.

Crain, Stephen, and Mark Steedman. 1985. On not being led up the garden path: The use of context by the psychological syntax processor. In *Natural language parsing: Psychological, computational, and theoretical perspectives*, ed. David R. Dowty, Lauri Karttunen, and Arnold M. Zwicky, 320–358. Cambridge: Cambridge University Press.

Davidson, Donald. 1974. Belief and the basis of meaning. *Synthese* 27:309–323.

Dehaene, Stanislas. 1997. *The number sense: How the mind creates mathematics*. Oxford: Oxford University Press.

Dixit, Avinash, and Barry Nalebuff. 1991. *Thinking strategically: The competitive edge in business, politics, and everyday life*. New York: W. W. Norton.

———. 2008. *The art of strategy: A game theorist's guide to success in business and life*. New York: W. W. Norton.

Dixit, Avinash, and Susan Skeath. 2004. *Games of strategy*. 2d ed. New York: W. W. Norton.

Dodge, Robert. 2006. *The strategist: The life and times of Thomas Schelling*. Hollis, NH: Hollis Publishing.

Dummett, Michael. 1981. *Frege: Philosophy of language*. 2d ed. Cambridge, MA: Harvard University Press.

Epstein, Joshua M. 2006. *Generative social science: Studies in agent-based computational modeling*. Princeton, NJ: Princeton University Press.

Evans, Gareth. 1982. *Varieties of reference*. Edited by John McDowell. New York: Oxford University Press.

———. 1985. *Collected papers*. Edited by Antonia Philips. Oxford: Oxford University Press.

Fagin, Ronald, Joseph Y. Halpern, Yoram Moses, and Mushe Y. Vardi. 1995. *Reasoning about knowledge*. 2d ed. Cambridge, MA: MIT Press.

Ferrand, Ludovic, and Boris New. 2003. Semantic and associative priming in the mental lexicon. In *Mental lexicon: Some words to talk about words*, ed. Patrick Bonin, 25–43. Hauppauge, NY: Nova Science Publishers.

Fodor, Jerry. 1970. Three reasons for not deriving "kill" from "cause to die". *Linguistic Inquiry* 1(4):429–438.

———. 1975. *The language of thought*. Cambridge, MA: Harvard University Press.

———. 1980. Methodological solipsism considered as a research strategy in cognitive psychology. *Behavioral and Brain Sciences* 3(1):63–73. Reprinted in *The Philosophy of Science*, ed. Richard Boyd, Philip Gasper and J. D. Trout, 651–670. Cambridge, MA: MIT Press, 1991.

Frank, Robert. 2002. *Phrase structure composition and syntactic dependencies*. Cambridge, MA: MIT Press.

Gärdenfors, Peter. 2000. *Conceptual spaces: The geometry of thought*. Cambridge, MA: MIT Press.

Gintis, Herbert. 2000. *Game theory evolving: A problem-centered introduction to modeling strategic interaction*. Princeton, NJ: Princeton University Press.

Glimcher, Paul W. 2003. *Decisions, uncertainty, and the brain: The science of neuroeconomics*. Cambridge, MA.: MIT Press.

Glimcher, Paul W., Michael C. Dorris, and Hannah M. Bayer. 2005. Physiological utility theory and the neuroeconomics of choice. *Games and Economic Behavior* 52:213–256.

Goffman, Erving. 1959. *The presentation of self in everyday life*. New York: Anchor Books.

———. 1967. *Interaction ritual: Essays on face-to-face behavior*. New York: Pantheon.

Goodman, Nelson. 1977. *The structure of appearance*. Pallas Paperbacks. Dordrecht, Netherlands: D. Reidel.

———. 1979. *Fact, fiction, and forecast*. Cambridge, MA: Harvard University Press.

Grice, H. Paul. 1957. Meaning. *Philosophical Review* 66(3):377–388.

———. 1975. Logic and conversation. In *Syntax and semantics*. Vol. 3: *Speech Acts*, ed. P. Cole and J. P. Morgan, 41–58. New York: Academic Press.

———. 1989. *Studies in the ways of words*. Cambridge, MA: Harvard University Press.

Groenendijk, Jeroen, and Martin Stokhof. 1991. Dynamic predicate logic. *Linguistics and Philosophy* 14:39–100.

Grosz, Barbara, Aravind Joshi, and Scott Weinstein. 1983. Providing a unified account of definite noun phrases in discourse. In *Proceedings of the 21st annual meeting of the ACL*, 44–50.

———. 1986. Towards a computational theory of discourse interpretation. Unpublished manuscript.

Gupta, Anil, and Nuel Belnap. 1993. *The revision theory of truth*. Cambridge, MA: MIT Press.

Hale, Ken, and Samuel Jay Keyser. 2002. *Prolegomenon to a theory of argument structure*. Cambridge, MA: MIT Press.

Hamilton, William D. 1964. The genetical evolution of social behavior: I and II. *Journal of Theoretical Biology* 7:1–16; 17–52.

———. 1967. Extraordinary sex ratios. *Science* 156:477–488.

———. 2001. My intended burial and why. *Ethology Ecology and Evolution* 12:111–122. Originally published in 1991.

Harper, D.G.C. 1982. Competitive foraging in mallards: 'Ideal free' ducks. *Animal Behaviour* 30(2):575–584.

Harris, Roy. 1988. *Language, Saussure and Wittgenstein: How to play games with words*. London: Routledge.

Harsanyi, John C., and Reinhard Selten. 1988. *A general theory of equilibrium selection in games*. Cambridge, MA: MIT Press.

Hausner, M., J. Nash, L. Shapley, and M. Shubik. 1964. "So long sucker"—a four-person game. In *Game theory and related approaches to social behavior: Selections*, ed. Martin Shubik, 359–361. New York: Wiley.

Hintikka, Jaakko. 1996. *The principles of mathematics revisited*. Cambridge: Cambridge University Press.

Hintikka, Jaakko, and Jack Kulas. 1983. *The game of language: Studies in game-theoretical semantics and its applications*. Dordrecht, Netherlands: D. Reidel.

———. 1985. *Anaphora and definite descriptions: Two applications of game-theoretical semantics*. Dordrecht, Netherlands: D. Reidel.

Hintikka, Jaakko, and Gabriel Sandu. 1997. Game-theoretical semantics. In *Handbook of logic and language*, ed. Johan van Benthem and Alice ter Meulen, 361–410. Cambridge, MA: MIT Press.

Hobbes, Thomas. 1968. *Leviathan*, ed. C. B. Macpherson. Baltimore: Penguin Books. Originally published 1651.

Hofbauer, Josef, and Karl Sigmund. 1998. *Evolutionary games and population dynamics*. 2d ed. Cambridge: Cambridge University Press.

Horn, Laurence R. 2001. *A natural history of negation*. Stanford, CA: CSLI Publications.

Jackendoff, Ray S. 1983. *Semantics and cognition*. Cambridge, MA.: MIT Press.

Jäger, Gerhard. 2007. Evolutionary game theory and typology: A case study. *Language* 83:74–109.

Kandori, Michihiro, George J. Mailath, and Rafael Rob. 1993. Learning, mutation, and long-run equilibria in games. *Econometrica* 61(1):29–56.

Kaplan, Fred. 1984. *The wizards of Armageddon*. New York: Simon and Schuster.

Kot, Mark. 2001. *Elements of mathematical ecology*. Cambridge: Cambridge University Press.

Kripke, Saul A. 1977. Speaker's reference and semantic reference. In *Contemporary perspectives in the philosophy of language*, ed. P. A. French, T. E. Uehling, Jr., and H. K. Wettstein, 6–27. Minneapolis: University of Minnesota Press.

———. 1982. *Wittgenstein on rules and private language*. Cambridge, MA: Harvard University Press.

Kroch, Anthony. 1989. Function and grammar in the history of English: Periphrastic 'do'. In *Language variation and change: Current issues in linguistic theory*, vol. 52, ed. Ralph Fasold and Deborah Schiffrin, 133–172. Philadelphia: John Benjamins.

Labov, William. 1973. The boundaries of words and their meanings. In *New ways of analysing variation in English*, ed. C.-J. N. Bailey and R. W. Shuy, 340–373. Washington, DC: Georgetown University Press.

———. 1994. *Principles of linguistic change: Internal factors*. Oxford: Blackwell Publishers.

———. 2010. *Principles of linguistic change*. Vol. 3: *Cognitive and cultural factors*. Oxford: Wiley-Blackwell.

Lakoff, George, and Mark Johnson. 1980. *Metaphors we live by*. Chicago: University of Chicago Press.

Landman, Fred. 1991. *Structures for semantics*. Dordrecht, Netherlands: Kluwer.

Lee, Christopher J. 1990. Some hypotheses concerning the evolution of polysemous words. *Journal of Psycholinguistic Research* 19(4):211–219.

Levinson, Stephen C. 1983. *Pragmatics*. Cambridge Textbooks in Linguistics. Cambridge: Cambridge University Press.

———. 2000. *Presumptive meanings: The theory of generalized conversational implicature*. Cambridge, MA: MIT Press.

Lewis, David. 1969. *Convention: A philosophical study*. Cambridge, MA: Harvard University Press.

———. 1983. Scorekeeping in a language game. In *Philosophical papers*, vol. I, 233–249. New York: Oxford University Press.

Littlewood, John E. 1953. *A mathematician's miscellany*. London: Methuen.

Luce, R. Duncan, and Howard Raiffa. 1957. *Games and decisions: Introduction and critical survey*. Wiley. Reprinted by Dover Publications, 1989.

Lyons, John. 1977. *Semantics*. 2 vols. Cambridge: Cambridge University Press.

Mackay, Charles. 1969. *Extraordinary popular delusions and the madness of crowds*. New York: Noonday Press.

Macrae, Norman. 1992. *John von Neumann*. New York: Pantheon.

May, Robert M. 1973. *Stability and complexity in model ecosystems*. Princeton, NJ: Princeton University Press.

Maynard Smith, John. 1972. *On evolution*. Edinburgh: Edinburgh University Press.

———. 1982. *Evolution and the theory of games*. Cambridge: Cambridge University Press.

Mayol, Laia. 2009. Pronouns in Catalan: Information, discourse and strategy. Ph.D. diss., University of Pennsylvania, Philadelphia.

Mayol, Laia, and Robin Clark. 2010. Overt pronouns in Catalan: Games of partial information and the use of linguistic resources. *Journal of Pragmatics* 42:781–799.

McCarthy, Cormac. 2005. *No country for old men*. New York: Random House.

McElreath, Richard, and Robert Boyd. 2007. *Mathematical models of social evolution*. Chicago: University of Chicago Press.

McNamara, Robert S., James G. Blight, Robert K. Brigham, with Thomas J. Biersteker and Herbert Y. Schandler. 1999. *Argument without end: In search of answers to the Vietnam tragedy*. New York: Public Affairs.

Mehta, Judith, Chris Starmer, and Robert Sugden. 1994. The nature of salience: An experimental investigation of pure coordination games. *American Economic Review* 84(3):658–673.

Miller, John H., and Scott E. Page. 2007. *Complex adaptive systems: An introduction to computational models of social life*. Princeton, NJ: Princeton University Press.

Mishkin, Frederic S. 1989. *The economics of money, banking, and financial markets*. 2d ed. Glenview, IL: Scott, Foresman.

Montague, Richard. 1974. *Formal philosophy: Selected papers of Richard Montague*. Edited and with an introduction by Richmond H. Thomason. New Haven, CT: Yale University Press.

Murphy, Gregory L. 2002. *The big book of concepts*. Cambridge, MA: MIT Press.

Muskens, Reinhard, Johan van Benthem, and Albert Visser. 1997. Dynamics. In *Handbook of logic and language*, ed. Johan van Benthem and Alice ter Meulen, 587–648. Cambridge, MA: MIT Press.

Myerson, Roger B. 1991. *Game theory: Analysis of conflict*. Cambridge, MA: Harvard University Press.

Neale, Stephen. 1990. *Descriptions*. Cambridge, MA: MIT Press.

Noë, Alva. 2009. *Out of our heads: Why you are not your brain, and other lessons from the biology of consciousness*. New York: Hill and Wang.

Norris, Dennis. 2006. The Bayesian reader: Explaining word recognition as an optimal Bayesian decision process. *Psychological Review* 113(2):327–357.

———. 2009. Putting it all together: A unified account of word recognition and reaction-time distributions. *Psychological Review* 116(1):207–219.

Nowak, Martin A. 2006. *Evolutionary dynamics: Exploring the equations of life*. Cambridge, MA: Belknap Press.

Nowak, Martin A., Joshua B. Plotkin, and David C. Krakauer. 1999. The evolutionary language game. *Journal of Theoretical Biology* 200:147–162.

Osborne, Martin J. 2004. *An introduction to game theory*. Oxford: Oxford University Press.

Osborne, Martin J., and Ariel Rubinstein. 1994. *A course in game theory*. Cambridge, MA: MIT Press.

Parikh, Prashant. 2001. *The use of language*. Stanford, CA: CSLI Publications.

———. 2010. *Equilibrium semantics*. Cambridge, MA: MIT Press.

Parikh, Prashant, and Robin Clark. 2007. An introduction to equilibrium semantics for natural language. In *Game theory and linguistic meaning*, ed. Ahti-Veikko Pietarinen. Vol. 18 of Current Research in the Semantics/Pragmatics Interface. Amsterdam: Elsevier.

Partee, Barbara H., Alice ter Meulen, and Robert E. Wall. 1990. *Mathematical methods in linguistics*. Dordrecht, Netherlands: Kluwer.

Pessin, Andrew, and Sanford Goldberg, eds. 1996. *The twin earth chronicles: Twenty years of reflection on Hilary Putnam's "The Meaning of 'Meaning'"*. Armonk, NY: M. E. Sharpe.

Pietarinen, Ahti-Veikko. 2003. Games as formal tools versus games as explanations in logic and science. *Foundations of Science* 8:317–364.

———. 2007. Semantic games and generalised quantifiers. In *Game theory and linguistic meaning*, ed. Ahti-Veikko Pietarinen, 183–206. Vol. 18 of Current Research in the Semantics/Pragmatics Interface. Amsterdam: Elsevier.

———. 2008. Who plays games in philosophy? In *Philosophy looks at chess*, ed. Benjamin Hale, 119–136. Chicago: Open Court.

Pietarinen, Ahti-Veikko, ed. 2007. *Game theory and linguistic meaning*. Vol. 18 of Current Research in the Semantics/Pragmatics Interface. Amsterdam: Elsevier.

Pietroski, Paul. 2005. Meaning before truth. In *Contextualism in philosophy*, ed. Gerhard Preyer and Georg Peter, 253–300. Oxford: Oxford University Press.

Pinker, Steven. 1994. *The language instinct: How the mind creates language*. New York: William Morrow.

———. 2007. *The stuff of thought: Language as a window into human nature*. New York: Viking.

Plato. 1945. *The Republic of Plato*. Translated with introduction and notes by Francis MacDonald Cornford. Oxford: Oxford University Press. Originally written ca. 380 B.C.E.

———. 1961. Cratylus. In *The collected dialogues of Plato including the letters*, ed. Edith Hamilton and Huntington Cairns, 421–474. Princeton, NJ: Princeton University Press. Originally written ca. 360 B.C.E.

Posner, M. I., and S. W. Keele. 1968. On the genesis of abstract ideas. *Journal of Experimental Psychology* 77:353–363.

———. 1970. Retention of abstract ideas. *Journal of Experimental Psychology* 83:304–308.

Poundstone, William. 1992. *Prisoner's dilemma*. New York: Anchor Books.

Putnam, Hilary. 1975. The meaning of "meaning". In *Language, mind, and knowledge*, ed. Keith Gunderson, 131–193. Minneapolis: University of Minnesota Press.

———. 1981. *Reason, truth and history*. Cambridge: Cambridge University Press.

Pylkkänen, Liina, Rodolfo Llinás, and Gregory L. Murphy. 2006. The representation of polysemy: MEG evidence. *Journal of Cognitive Neuroscience* 18(1):97–109.

Quine, W.V.O. 1960. *Word and object.* Cambridge, MA: MIT Press.

Rorty, Richard. 1979. *Philosophy and the mirror of nature.* Princeton, NJ: Princeton University Press.

Rosch, Eleanor. 1975. Cognitive reference points. *Cognitive Psychology* 7(4):532–547.

———. 1978. Principles of categorization. In *Cognition and categorization,* ed. E. Rosch and B. B. Lloyd, 27–48. Hillsdale, NJ: Lawrence Erlbaum.

Rosch, Eleanor, and Carolyn B. Mervis. 1975. Family resemblances: Studies in the internal structure of categories. *Cognitive Psychology* 7:573–605.

Ross, Ian. 2006. Games interlocutors play: New adventures in compositionality and conversational implicature. Ph.D. diss., University of Pennsylvania, Philadelphia.

Rousseau, Jean Jacques. 2004. *Discourse on inequality.* Translated by G.D.H. Cole. Whitefish, MT: Kessinger Publishing. Originally written in 1754.

Rubinstein, Ariel. 1998. *Modeling bounded rationality.* Cambridge, MA: MIT Press.

———. 2000. *Language and economics: Five essays.* Cambridge: Cambridge University Press.

Sally, David. 2002. What an ugly baby! Risk dominance, sympathy, and the coordination of meaning. *Rationality and Society* 14(1):78–108.

———. 2003. Risky speech: Behavioral game theory and pragmatics. *Journal of Pragmatics* 35:1223–1245.

Saussure, Ferdinand de. 1966. *Course in general linguistics.* Translated from the French by Wade Baskin. New York: McGraw-Hill. Originally published 1916.

Schelling, Thomas C. 1960. *The strategy of conflict.* Cambridge, MA: Harvard University Press.

———. 1978. *Micromotives and macrobehavior.* New York: W. W. Norton.

———. 2006. *Strategies of commitment and other essays.* Cambridge, MA: Harvard University Press.

Searle, John. 1969. *Speech acts: An essay in the philosophy of language.* Cambridge: Cambridge University Press.

———. 1984. *Minds, brains and science.* Cambridge, MA: Harvard University Press.

———. 1995. *The construction of social reality.* New York: Free Press.

Sevenster, Merlijn. 2006. Branches of imperfect information: Logic, games, and computation. Ph.D. diss., University of Amsterdam.

Shoham, Yoav, and Kevin Leyton-Brown. 2009. *Multiagent systems: Algorithmic, game-theoretic, and logical foundations.* Cambridge: Cambridge University Press.

Sigmund, Karl. 1993. *Games of life: Explorations in ecology, evolution and behavior*. Oxford: Oxford University Press.

Simon, Herbert A. 1955. A behavioral model of rational choice. *Quarterly Journal of Economics* 69(1):99–118.

———. 1956. Rational choice and the structure of the environment. *Psychological Review* 63(2):129–138.

———. 1982. *Models of bounded rationality: Behavioral economics and business organization*. Cambridge, MA: MIT Press.

Simpson, Greg B. 1994. Context and the processing of ambiguous words. In *Handbook of psycholinguistics*, ed. Morton Ann Gernsbacher, 359–374. San Diego: Academic Press.

Skyrms, Brian. 2004. *The stag hunt and the evolution of social structure*. Cambridge: Cambridge University Press.

Smullyan, Raymond M. 2009. *Logical labyrinths*. Wellesley, MA: A. K. Peters.

Soames, Scott. 2003. *Philosophical analysis in the twentieth century*. Vol. I: *The Dawn of Analysis*. Princeton, NJ: Princeton University Press.

Strawson, P. F. 1950. On referring. *Mind* 59:320–344.

Sugden, Robert. 1995. A theory of focal points. *Economic Journal* 105(430):533–550.

Sugden, Robert, and Ignacio E. Zamarrón. 2006. Finding the key: The riddle of focal points. *Journal of Economic Psychology* 27(5):609–621.

Szymanik, Jakub. 2009. Quantifiers in time and space: Computational complexity of generalized quantifiers in natural language. Ph.D. diss., University of Amsterdam.

Talmy, Leonard. 1985. Lexicalization patterns: Semantic structure in lexical forms. In *Language typology and syntactic description*. Vol. 3: *Grammatical categories and the lexicon*, ed. Timothy Shopen, 57–149. Cambridge: Cambridge University Press.

Tarski, Alfred. 1983. The concept of truth in formalized languages. In *Logic, semantics, metamathematics*, 2d ed., ed. John Corcoran, trans. J. H. Woodger, 152–278. Indianapolis: Hackett.

Taylor, John R. 1995. *Linguistic categorization: Prototypes in linguistic theory*. 2d ed. Oxford: Oxford University Press.

Tomalin, Marcus. 2006. *Linguistics and the formal sciences: The origins of generative grammar*. Cambridge: Cambridge University Press.

Turing, Alan. 1950. Computing machinery and intelligence. *Mind* 59:433–460. Also in *Computation and intelligence: Collected readings*, ed. George F. Luger, 23–46. Cambridge, MA: MIT Press, 1995.

Tversky, Amos. 1977. Features of similarity. *Psychological Review* 84(4):327–352.

van Benthem, Johan. 2011. *Logical dynamics of information and interaction*. Cambridge: Cambridge University Press.

van Deemter, Kees. 2010. *Not exactly: In praise of vagueness*. Oxford: Oxford University Press.

van Rooy, Robert. 2003. Being polite is a handicap: Towards a game theoretical analysis of polite linguistic behavior. In *Tark '03: Proceedings of the 9th ACM conference on theoretical aspects of rationality and knowledge*, 45–58.

von Neumann, John. 1957. Passing of a great mind, by Clay Blair, Jr. Obituary. *Life*, February 25, 1957, 89+.

von Neumann, John, and Oskar Morgenstern. 1944. *Theory of games and economic behavior*. Princeton, NJ: Princeton University Press.

Walker, Marilyn, Aravind Joshi, and Ellen Prince, eds. 1998. *Centering theory in discourse*. Oxford: Clarendon Press.

Walker, Marilyn, and Ellen Prince. 1996. A bilateral approach to givenness: A hearer-status algorithm and a centering algorithm. In *Reference and referent accessibility*, ed. Thorstein Fretheim and Jeanette Gundel, 291–306. Philadelphia: John Benjamins.

Wall, Robert. 1972. *Introduction to mathematical linguistics*. Englewood Cliffs, NJ: Prentice Hall.

Williamson, Timothy. 1994. *Vagueness*. London: Routledge.

Wittgenstein, Ludwig. 1953. *Philosophical investigations*. Edited by G.E.M. Anscombe and R. Rhees. Translated by G.E.M. Anscombe. Malden, MA: Blackwell.

———. 1958. *The blue and brown books*. New York: Harper and Row. Original notes dictated in English 1933–1935.

Young, H. Peyton. 1993. The evolution of conventions. *Econometrica* 61(1):57–84.

———. 1998. *Individual strategy and social structure: An evolutionary theory of institutions*. Princeton, NJ: Princeton University Press.

Zahavi, Amotz. 1975. Mate selection—a selection for a handicap. *Journal of Theoretical Biology* 53:205–214.

Zelizer, Viviana A. 1997. *The social meaning of money: Pin money, paychecks, poor relief, and other currencies*. Princeton, NJ: Princeton University Press.

Zipf, George K. 1949. *Human behavior and the principle of least-effort*. New York: Addison-Wesley.

Index

Abélard, Peter, 128–129
Abélard and Eloïse. *See* Verifier and falsifier
Accommodation, 206–207, 213
 and common knowledge, 211
 and coordination, 210
 and game-theoretic semantics, 209, 210
 and presupposition, 204–211
 and social conventions, 207
Accusative. *See* Case-marking
Action and rational choice, 68
Agent-based modeling, 323–326
Allegory of the Cave, 28–30, 42
Altruism, 118
Ambiguity, 215, 315
 and coordination, 218
 and expected utility, 228–229
 and objective probability, 238
 and semantic hierarchies, 309–311
 and subjective probability, 229–231
Appropriateness conditions, 284, 287
Appropriateness of usage. *See* Appropriateness conditions
Aristotelian definition, 327
 and set extensions, 284, 285
 and vagueness, 286
Aristotelian Square of Opposition, 150–153
 and contradictories, 153
 and contraries, 153
 and entailment, 153
 and subcontraries, 153
 and syllogisms, 150

Arms race, 102
Armstrong, S., 288–289, 327
Assurance games, 330
Atomic sentences. *See also* Game-theoretic semantics
 and truth definitions, 139
Attractors, 118
 and evolutionary game theory, 118
Austin, J. L., xiii, 59, 287, 328
Axelrod, R., 96, 123

Bach or Stravinsky. *See* Game, Hobo Dinner
Backward induction (rollback), 75
Bardsley, N., 327
Barwise, J., 19, 42, 148, 177, 178, 181
Basic level terms, 312
Battle of the sexes. *See* Game, Hobo Dinner
Beauty contest game, 192
 and bounded rationality, 192–193
Beaver, D., 213
Belnap, N., 18
Benz. A., 122
Bever, T., 243
Binmore, K., 122
Blackburn, S., 18
Borel, É., 122
Bostock, D., 42
Bound pronouns, 161–162. *See also* Game-theoretic semantics
Bounded rationality, 191, 212
 and common knowledge, 191
Bourdieu, P., 189

Boyd, R., 123
Breheny, R., 280
Britton, B., 320
Brown, P., 40, 267, 268, 274, 277, 278, 281, 331
Bryan, M., 52
Bundy, M., 65

Cake game, 69–75
Camerer, C., 193, 212
Carlson, L., 177
Carnap, R., 40
Carroll, L., 43
Carrying capacity
 and competition, 304
 and lexical competition, 305
Case-marking
 accusative system, 113
 ergative system, 113
 and evolutionary game theory, 112–118
 split ergative, 117
Categorization and prototype theory, 291
Category resemblance, 311–312
Centering theory, 280
 and prominence hierarchy, 253
Central tendency and prototype theory, 289–290, 292
Cheap talk systems, 189
Chicken game and evolutionary game theory, 109–110
Chinese room, 36–38, 42
Choice functions, 168
 and best response, 168
Chomsky, N., 40
CHRISTINE corpus, 116
Clark, H., 183, 184, 187, 194, 197, 198, 209, 212
Clark, R., 19, 178, 242, 243 245, 247, 252, 280, 380
Clinton, W., 88
Cognitive economy and prototype theory, 287
Common knowledge, 46–50, 266
 and community membership, 194–195

and game theory, 49
and public information, 181
sources of, 198
Communication
 backchannel, 204–211
 and bounded rationality, 204
 and tacit bargaining, 295
Communicative economy, 246
Competition coefficients, 304
Competition model, 303–306
Completeness, 33–34
Compositionality, 15, 19
 and game-theoretic semantics, 174
 and grammar, 23–25
 and Mentalese, 16–18
Computational theory of mind, 26–27
Concepts, 10, 11, 289, 327
 and essentialism, 11
 and informational categories, 10–11
 and number sense, 12
 and social behavior, 301
Conditioned linguistic rule, 89
Consequences (of action), 68
Consistency, 33–34
Context Free Phrase Structure Grammar (CFPSG), 21–23
Contradictories. See Aristotelian Square of Opposition
Contraries. See Aristotelian Square of Opposition
Conventions, 236
 evolution of, 123
Conversational implicature. See Implicature
Conversational maxims, 224, 245, 265, 268, 276
 and face, 273
 and games of partial information, 264
 manner, 223
 quantity, 223–224
 and utility, 225
Cooperation, 330
 evolution of, 97–98, 278
 and linguistic conventions, 97
 and linguistic meaning, 177

Cooperative principle, 223
Coordinated attack, 182–183, 212
  and common knowledge, 183
Coordination. *See also* Coordination game
  and focal points, 295–296
Coordination game, 89–105, 326
  and mixed strategy, 91–92
  and reference, 92
  and signaling, 90
  and vagueness, 300
Copresence
  indirect, 197–198
  linguistic, 195–197
  physical, 195
Correspondence theory of meaning, 176
Costs and language production, 252
Cowie, F., 18
Crain, S., 243
Cratylus, 14–16
Cue validity, 311–312, 328

Dahl, Ö, 116
Davidson, D., 18
Definite descriptions, 183–184, 213
  and expected utility, 232–234
  and games of partial information, 232–234
  and mutual knowledge, 184–191
  and presupposition, 205–206
  Russellian treatment of, 205
  and strategic decision-making, 248–250
Dehaene, S., 19
Dirty frat boy problem, 46–48
  and rationality, 192
  sources of, 59
Discourse anaphora, 152, 245
  and contrastive stress, 261
  and games of partial information, 245–263
  and game-theoretic semantics, 246
  and multiple pronouns, 257–263
  and Nash equilibrium, 256, 260–261
  and production costs, 254
  and rational choice, 263
  and sequential games, 263
  and strategic decision-making, 247
Dixit, A., 75, 122
Downward entailment, 173–174
Ducks, ideal free, 85–87, 122
Dummett, M., 19

Ecology and game theory, 58–59
Economy principles, 113
Elementary trees
  and intransitive verbs, 136
  and logical operators, 136
  and nouns, 135–136
  and transitive verbs, 136–137
Eloïse, 128–129
Emergent behavior, 87
  and meaning, 285
Entailment, 9
  and Aristotelian Square of Opposition, 153
  and game-theoretic semantics, 173
  syllogism, 10
Entrenchment and basic level terms, 314
Epstein, J., 328
Equilibrium selection, 100–101, 234–237
Equilibrium strategy, 71
  and frequency dependence, 109
  and rollback equilibrium, 75
  and Stag Hunt, 99–100
  and utility, 71
Ergative. *See* Case-marking
ESS. *See* Strategy
Etchemendy, J., 19, 42, 177, 178
Euclid, 40–41
Evans, G., 213
Evolution of convention, 322
Evolutionary game theory, 106–122. *See also* Game
  and ESS, 323
  and face, 278–279
  and iterated games, 108
  and mixed strategy, 109
  and Stag Hunt game, 111
Existential sentences. *See* Game-theoretic semantics

Expected utility. *See* Utility
Exploitation competition, 331
Extensive normal form, 73. *See also* Game tree

Face, 267–268
 and common knowledge, 270
 face threatening act, 259–270
 negative, 268–269
 positive, 268–269
 as social construction, 268
 and social hierarchies, 270–272
 threats to, 269
 and utility, 272
Fagin, R., 183, 212
Faithfulness, 116
Falsification and game-theoretic semantics, 127
Ferrand, L., 243
Fey, T., 89
Focal interpretation, 297
 and metrics, 300
Focal point, 42, 196, 292–300
 and concepts, 302
 and coordination games, 293–294
 and discourse anaphora, 253–254, 259, 262–263
 and primary salience, 294
 and secondary salience, 294
 and semantic space, 298, 326
Fodor, J., 18, 19, 26, 27, 41
Formality condition. *See* Methodological solipsism
Frank, R., 178
Frege, G., 19
Fregean compositionality. *See* Compositionality
Freud, S., 7
FTA (Face Threatening Act). *See* Face
Fuck Your Buddy, 66–67
Fuzzy categories and prototype theory, 288

Game
 Bach or Stravinsky, 330
 Battle of the Sexes, 330
 Cake, 69–75
 Chicken, 109 (*see also* Game, sidewalk)
 evolutionary, 106–118
 foraging, 85–87
 Garden, 75–79
 Hawk-Dove, 106–111, 278–279, 323
 Hobo Dinner, 90–91, 123
 Holmes-Moriarty, 80–85, 122, 220
 and incomplete information, 74, 75
 and partial information, 75
 and perfect information, 74
 Prisoner's Dilemma, 93–95, 278, 323, 330
 rock-paper-scissors, 81–82
 sidewalk, 95–96, 98 (*see also* Game, Chicken)
 Signaling, 90
 Stag Hunt, 98–100, 123, 267
 Stop Sign, 96
 strategic, 83
Game matrix (strategic normal form), 80–81
Game of partial information, 219
Game-theoretic semantics, 127
 and ambiguity, 172
 and appropriateness conditions, 301
 and atomic sentences, 137–138
 and bound pronouns, 161–162
 and cardinals, 169–172
 and choice functions, 168
 and compositionality, 174
 and conjunction, 142–144
 and correspondence theory of meaning, 176
 and discourse anaphora, 152, 154, 157
 and disjunction, 145–146
 and entailment, 173
 and equilibrium, 144
 and existential sentences, 153–157
 and material implication, 147–150
 and mereology, 158–159
 and model theory, 137
 and monotonicity, 173–174
 and negation, 140–141
 and scope, 162–169
 and truth, 137
 and universal sentences, 159–161

and vagueness, 301
and zero-sum games, 145
Game theory and rational choice, 67
Game tree, 70
Garden path sentences, 237–242
Gärdenfors, P., 298
Gause's principle. *See* Principle of competitive exclusion
Gleitman, H., 288–289, 327
Gleitman, L., 288–289, 327, 329
Glimcher, P., 122
Goffman, E., 267, 281
Goldberg, S., 42
Goodman, N., 40, 313–314, 328
Grammar
  and parse trees, 23
  and rules, 22
  use of, 279–280
Grammatical functions and prominence, 253
Grice, H. Paul, xii, xiii, 213, 222, 223, 242, 245, 264, 267, 280
Grossman, M., 19
Grosz, B., 280
Gupta, A., 18

Haldane, J.B.S., 119
Hale, K., 19
Halpern, J., 183, 212
Hamilton, W. D., 119, 121, 122, 123
Hamilton's rule, 119–120
Harper, D.G.C., 85
Harris, R., 59
Harsanyi, J., 123
Hausner, M., 67
Hearer economy. *See* Economy principles
Hintikka, J., 127, 175, 177, 178
Hobbes, T., 95
Hofbauer, J., 123, 328
Homophony, 315
  as historical accident, 316
  representation of, 318
Horn, L., 242, 281
Hypothesis of the Universality of the Division of Linguistic Labor, 57, 58, 212

Imitation dynamics, 117
Implicature, 103–104
  cancellation, 103, 225
  and cooperation, 267
  and equilibrium selection, 234–237
  and face threatening acts, 273
  and irony, 235–236
  and politeness, 263–280
  and truth conditions, 224–225
Inclusive fitness, 121
Incomplete information, 74. *See also* Game
Independence-friendly logic, 145, 178
Indirect copresence, 197–198
Indirectness and risk, 267
Informational categories, 11, 12
Information set, 216
  and lexical access, 217–218
Intentionality, 37–38, 39
  social nature of, 38–39
Intentions
  communicative, 221–222
  and information states, 221
Interference competition, 303
Irony, 235–236
  and equilibrium selection, 235–236
  and face, 277–278

Jackendoff, R., 19
Jäger, G., 59, 112, 113, 116, 122
Johnson, L. B., 65
Johnson, M., 327
Joshi, A., 280

Kandori, M., 123
Kaplan, F., 65
Keele, S., 291
Keyser, S. J., 19
Kot, M., 328
Krakauer, D., 59
Kripke, S., 42, 59, 316–317
Kroch, A., 123
Kulas, J., 177

Labov, W., 123, 201, 202, 203, 213, 291, 298, 300, 328, 331
Lakoff, G., 327

Landman, F., 42
Language
  as economic system, 189
  rational use of, 223
Language of Thought. *See* Mentalese
Language use and rational choice, 247
Lee, C. J., 320
Levinson, S., 40, 59, 242, 267, 268, 274, 277–278, 281, 331
Lewis, D., 294, 327
Lexical access
  and ambiguity, 215–221
  and conversational maxims, 225–226
  economic model of, 241
  and expected utility, 230–231
  and mixed strategy, 228
  and objective probability, 238
Lexical competition, 302–303
  and ecology, 303
  and Lotka-Volterra competition model, 305–307
Lexical decomposition, 24, 25, 30
Lexical frequency and garden path sentences, 239
Lexical semantics and discourse anaphora, 262
Liar paradox, 6, 18
Linguistic copresence, 195–197
  potential, 197
Linguistic defaults
  evolution of, 322
  and semantic hierarchies, 311
Linguistic signs
  conventional, 14
  natural, 14
Littlewood, J. E., 59
Llinás, R., 317–318, 321, 328
Logic. *See also* Game-theoretic semantics
  and completeness, 33–34
  and consistency, 33–34
  first-order, 34, 125
  and incomplete information, 145
  inferencing, 16–18, 19
  and proof theory, 32–35, 42, 329
  and rational choice, 128
  and zero-sum games, 125

Logical positivism, 12
Lotka-Volterra competition model, 303–306, 328
  and basic level terms, 314
  and polysemy, 321
Lotka, A., 303
Luce, R. D., 122

Mackay, C., 328
Macrae, N., 122
MAD (Mutually Assured Destruction), 65
Magnetoencephalography (MEG), 318
Mailath, G., 123
May, R., 328
Mayol, L., 252, 280
McCarthy, C., 303
McElreath, R., 123
McNamara, R., 66
McNaughton, J., 65
Meaning
  and constructivism, 285–286
  and cooperation, 286
  and coordination games, 326–327
  ecology of, 58
  and emergent properties, 328
  and grammar use, 209
  and rational use of language, 279–280
  social construction of, 55–58
  and social systems, 285
Meaning postulate, 17–18, 26
Mehta, J., 294–296, 327
Mentalese, 8, 12, 13, 18, 19, 40, 45
  and common knowledge, 46, 49
  and concepts, 10
  and evolution, 16
  and informational categories, 12
  and language variation, 44–45
  and mental representations, 14
  and Platonic heaven, 30
  and prototype theory, 289
  and translation theory, 12
  and vagueness, 300
Mereology, 155
  and plurals, 155
Mervis, C., 327

Methodological solipsism, 27, 28, 41, 57
 and capitalism, 30–31
Metonymy, 310, 325
 and semantic hierarchies, 313
Metrics and tacit bargaining, 300
Miller, J., 328
Miscommunication, 199–204, 253
 and bound pronouns, 199
 and coordination, 208
 and discourse anaphora, 253–254
 and implicature, 267
 phonological, 200–203
 and quantifier scope, 199–200
 rate of, 203
 and regional dialects, 202–203
Mishkin, F., 59
Mixed strategy, 74. *See also* Strategy profile
 and linguistic behavior, 88–89
Mixed strategy Nash equilibrium. *See* Strategy profile
Model theory, 127
Monotonicity and game-theoretic semantics, 173
Montague, R., 127, 327
Montague grammar, 41–42
Morgenstern, O., 122
Moses, Y., 183, 212
Moss, L., 181
Muddy children problem. *See* Dirty frat boy problem
Murphy, G., 42, 317–318, 321, 327, 328
Mutual knowledge, 183, 186. *See also* Common knowledge
 and communication, 187
 definition of, 187
Mutual knowledge paradox, 188
 and common knowledge, 188
 and coordinated attack, 191
Myerson, R., 59, 122

Nalebuff, B., 122
Nash equilibrium, 234
 and truth, 145
Nash, J., 66, 67

Nativism, 18
Neale, S., 213
Nerval, G., 56
Neumann, J. von, 64, 122
New, B., 243
Noë, A., 42
Noncompositional quantifiers, 175
Non-wellfounded sets, 181
Norris, D., 243
Nowak, M., 59, 123, 328
Number sense, 12, 19
Nunberg, G., 88

Off-record face threats, 275
O'Keefe, D., 52
On-record face threats, 275
Osborne, R., 122, 123

Page, S., 328
Pareto-dominance. *See* Payoff dominance
Parikh, P., xv, 123, 178, 242, 243, 245, 247, 280, 328
Parse trees. *See* Grammar
Partee, B., 41
Partial information, 75
 games of, 245
Payoff dominance, 92, 100, 123, 330
 and discourse anaphora, 261
 and equilibrium selection, 234
Perceived world structure and prototype theory, 287
Perfect information, 74, 219–220. *See also* Game
Perry, J., 178
Pessin, A., 42
Phonetic decision-making, 201–202
Physical copresence
 and common knowledge, 195
 potential, 195
Pietarinen, A.-V., 177, 178
Pietroski, P., 328
Pinker, S., 19
Plato, 28–30
Platonic heaven, 29
Platonism, 13–16
Plotkin, J., 59

Politeness
  formulas for, 272–273
  and games of partial information, 263–280
  and implicature, 245
  and linguistic conventions, 237
Polysemy
  compared to homophony, 315
  and coordination games, 321
  and diachrony, 320
  and ecology, 321
  evolution of, 322
  and focal points, 319
  and metrics, 292
  representation of, 318
  and semantic landscapes, 319
  and tacit bargaining, 320
Pop-out effects, 252, 280, 283
Posner, M., 291
Preference relation
  algebra of, 69
  and rational choice, 68
  and utility function, 76
Preferences, linguistic, 105
Presupposition, 213
  definition of, 205
  and game-theoretic semantics, 208–209
Primary salience and focal points, 294
Prince, E., 280
Principle of competitive exclusion, 307, 328
  and semantic niches, 308
Prisoner's Dilemma, 93–95, 123. *See also* Game
  and Hawk-Dove, 107–108
  iterated, 96–97
  and Tit for Tat, 96–97
Probability, 226–228
  and garden path sentences, 237–238
  objective, 238
  pronouns and gender, 248
  subjective, 229
  and utility, 83
Pronouns
  and definite descriptions, 248
  and probability, 248–249
  and strategic decision-making, 248–250
  and utility function, 251–252
Proof theory. *See* Logic
Prototypes, 287
  and Aristotelian properties, 288
  as conventionalized focal points, 297
  and focal points, 296
  and mathematical categories, 288–289
  and measurement, 288
  and metrics, 290
  and salient properties, 290
Public knowledge. *See* Common knowledge
Pure strategy, 74. *See* Strategy profile
Putnam, H., 41, 42, 56–57, 59. 212. 289
Pylkkänen, L., 317–318, 321, 328

Quine, W.V.O., 7, 18

Rai. *See* Stone money
Raiffa, H., 122
RAND Corporation, 63
Rational agent, 67, 191. *See also* Rational choice
  as decision maker, 72
  vs. nonrational agent, 106
Rational choice, defined, 68
Ray, S., 330
Reciprocity and evolution, 118–120
Red Queen hypothesis, 121
Reference and coordination games, 299, 326
Relevance theory, 280
Reynolds, M., 63
Risk dominance, 92, 93, 101
  and equilibrium selection, 234
  and implicature, 105
Rob, R., 123
Rollback. *See* Backward induction
Rollback equilibrium. *See* Equilibrium strategy
Rorty, R., 42
Rosch, E., 42, 312, 327
Rousseau, J.-J., 99

Rubinstein, A., 122, 194, 212, 213
Russell, B., 162, 205–206
Ryant, N., 243, 328

Sally, D., 123, 236, 242, 327
Sampson, G., 116
Samtal i Göteborg corpus, 116
Sandu, G., 177
Sankoff, G., 331
Satisficing, 182, 192
Saussure, F., 43–44, 51, 59
Schelling, T., xvi, 42, 65, 66, 122, 196, 292, 293, 294, 296, 301, 327, 328
Schelling point. *See* Focal point
Schelling salience, 294
Scope and pronouns, 166
Scope phenomena. *See* Game-theoretic semantics
Searle, J., 36, 37, 42, 59, 327
Secondary salience and focal points, 294
Self-evidence, 194
 and bounded rationality, 198–199
Selten, R., 123
Semantic hierarchies, 309–314
Semantic niche, 305
Semantic space, 298, 301–302
 optimal coverage of, 302
Semantics, truth-conditional, 127
Semenya, C., 289
Set theory and word meaning, 283–284
Sevenster, M., 178
Shapley, L., 67
Shared root hypothesis and polysemy, 317–318
Shubik, M., 67
Sidewalk game (Chicken), 95–96. *See also* Game
Sigmund, K., 123, 328
Signaling. *See* Coordination game
Simon, H., 191, 212
Simpson, G., 243
Situation semantics, 178
Skeath, S., 75, 122
Skolem functions. *See* Choice functions

Skyrms, B., 123
Smullyan, R., 18
Soames, S., 19
Solipsism, 212. *See also* Methodological solipsism
So Long Sucker, 66–67
Sorites paradox, 328
 and vagueness, 300
Spanish, 5–7
Speaker economy. *See* Economy principles
Starmer, C., 294–296, 327
Steedman, M., 243
Stevens, J., 331
Stochastically stable strategy, 118
Stone money, 50–53
Strategic game. *See* Game
Strategic normal form. *See* Game matrix
Strategy, evolutionarily stable (ESS), 107
Strategy profile, 71, 79
 and mixed strategy, 74, 83, 84
 and pure strategy, 74
Strawson, P. F., 205–206
Subcontraries. *See* Aristotelian Square of Opposition
Subgame perfection, 173
Substitution. *See* Tree-adjoining grammar (TAG)
Sugden, R., 294–296, 327
Syntax for semantic games, 129–136
Szymanik, J., 175, 178

Tacit bargaining, 295
 and reference, 299
Talmy, L., 58
Tarski, A., 18, 127, 327
Taylor, J., 327, 328
Ter Meulen, A., 41
Terminal history, 73
The Republic, 28, 42
Tit for Tat strategy, 330. *See also* Prisoner's Dilemma
 and ESS, 108
Topic heuristic, 209
 and cooperation, 209

Tree-adjoining grammar (TAG), 127, 178
  and adjunction, 132–135
  and semantic games, 129–136
  and substitution operation, 130–131
Truth
  correspondence theory of, 9
  and liar paradox, 6
  and Mentalese, 12
  predicate, 8
  and translation manuals, 7
  translation theory of, 5–6, 7, 9, 12, 13, 26
  and truth conditions, 13
Truth conditions, 125–126
Tucker, A., 330
Turing, A., 35, 42
Turing Machine, 27
Turing Test, 35–38, 42
  and intelligence, 35, 330n2
Twin Earth, 56, 330
Two general problem. *See* Coordinated attack

Universal sentences. *See* Game-theoretic semantics
Upward entailment, 173–174
Use-mention distinction, 6
Utility, 71, 122
  choice costs, 252
  communication factors, 252
  and communication games, 189–190, 222
  context factors, 252
  and conversational maxims, 252
  encoding costs, 252
  expected, 82, 86, 91, 115, 226
  and face, 274
  and face threatening acts, 275
  and foraging, 86
  hearer, 115
  and linguistic preferences, 279
  speaker, 115
  sympathy constant, 236, 242
  and utility function, 72

Vagueness, 286, 300
  and constructed terms, 286

Value, social construction of, 51, 54–55
Van Benthem, J., 178
Van Deemter, K., 286, 328
Van Rooy, R., 122, 281
Vardi, M., 183, 212
Variation and language, 88
Verification
  as game, 74
  and game-theoretic semantics, 127
Verifier and falsifier, 127, 128, 283, 301
  and communication games, 188
  and mereological sums, 158–159
  and truth definitions, 138
Vietnam, 65–66
Volterra, V., 303

Walker, M., 280
Wall, R., 40, 41
Weinstein, S., 280
Williamson, T., 328
Wittgenstein, L., xiii, 40, 42, 59, 328
Word sense and frequencies, 240

Xerox Corporation, 311, 313, 331

Yap, 59. *See also* Stone money
  as computational system, 53–55
Young, H.P., 118, 123

Zahavi, A., 81
Zahavi handicap principle, 281
Zamarrœn, I., 327
Zeevat, H., 213
Zelizer, V., 52, 53, 59
Zero-sum game. *See* Game
Zipf's Law, 322, 328, 332